Cambridge monographs in physical chemistry 1

GENERAL EDITORS PROFESSOR J. W. LINNETT F.R.S.
                 PROFESSOR J. H. PURNELL

Vibrational spectroscopy of solids

Vibrational spectroscopy of solids

P. M. A. SHERWOOD,
*Fellow of Downing College, Cambridge
and University Lecturer in Inorganic Chemistry,
University of Newcastle upon Tyne*

Cambridge at the University Press
1972

Published by the Syndics of the Cambridge University Press
Bentley House, 200 Euston Road, London NW1 2DB
American Branch: 32 East 57th Street, New York, N.Y. 10022

© Cambridge University Press 1972

Library of Congress Catalogue Card Number: 79-185566

ISBN: 0 521 08482 2

Printed in Great Britain
at the University Printing House, Cambridge
(Brooke Crutchley, University Printer)

TO MY PARENTS

Contents

Preface		*page* xi
1	Introduction	1
	1.1 The differences between the spectra of molecules in the solid-phase and molecules in the gas-phase	1
	1.2 Solid state vibrational spectra for which there is no gas-phase analogy	4
	1.3 The wave model for vibrational excitations in solids	4
2	Phonons and lattice vibrations	7
	2.1 Introduction	7
	2.2 Elastic waves	7
	2.3 Lattice dynamics of simple systems	7
	2.4 The finite lattice and boundary conditions	17
	2.5 The three-dimensional lattice, longitudinal and transverse waves	20
	2.6 External and internal vibrations	22
	2.7 The quantization of lattice vibrations	25
	2.8 Lattice dynamics of real crystals	27
	2.9 Experimental methods for determining dispersion relations	38
3	The application of group theory to a crystal lattice	46
	3.1 Introduction	46
	3.2 Groups	46
	3.3 The space group	48
	3.4 Subgroups of the finite space group	49
	3.5 Elements of the finite space group and its subgroups	52
	3.6 Space groups during lattice vibrations	54
	3.7 Unit cell group analysis	57
	3.8 Site group analysis	70
	3.9 Infrared and Raman selection rules	74
	3.10 Examples of the application of unit cell group analysis	79

Contents

4	The interaction of radiation with a crystal	page 83
	4.1 Introduction	83
	4.2 Models for studies of interaction mechanism	85
	4.3 Longitudinal-transverse splitting and the dielectric constant for an infinite cubic ionic crystal	85
	4.4 Reflection for an infinite ionic crystal	94
	4.5 Phonon–photon interaction and the polariton	95
	4.6 Absorption, reflection and the need for anharmonic forces	100
	4.7 The Raman scattering mechanism	104
	4.8 Raman scattering geometries	106
	4.9 General waves in vibrational spectra	109
	4.10 The effects of finite sample size	116
5	Second order vibrational spectroscopic features	136
	5.1 Introduction	136
	5.2 The width and temperature dependence of fundamentals	136
	5.3 Multiphonon processes	139
	5.4 Band shapes	151
	5.5 Internal modes	156
	5.6 External modes	162
	5.7 Particle scattering and sample interference	164
	5.8 Infrared intensities	166
	5.9 Raman intensities	171
	5.10 Resonant Raman scattering	174
	5.11 Brillouin scattering	174
	5.12 Pure absorption	175
	5.13 Attentuated total reflection	177
	5.14 Absorption by metals	180
	5.15 The choice of spectroscopic windows	182
	5.16 Pressure dependence of vibrational modes	182
	5.17 The effect of applied fields	184
	5.18 Hydrogen bonding	185
	5.19 Vibrational spectroscopic effects of crystal disorder and defects	187

Contents

6	Spectroscopic effects other than vibrational transitions in the vibrational energy region	*page* 195
6.1	Introduction	195
6.2	Excitons and electronic transitions	195
6.3	Plasmons	200
6.4	Magnons	202

Appendix: character tables		209
A.1	The C_s, C_i, and C_n groups	209
A.2	The D_n groups	210
A.3	The C_{nv} groups	210
A.4	The C_{nh} groups	211
A.5	The D_{nh} groups	212
A.6	The D_{nd} groups	213
A.7	The S_n groups	214
A.8	The cubic groups	215
A.9	The groups $C_{\infty v}$ and $D_{\infty h}$	216

References	217
Glossary of terms	237
Index	244

Preface

It is now routine for the chemist to record the spectrum of every new compound. For convenience, this is usually done on solid samples, rather than on solution or gas-phase samples. The chemist, however, frequently analyses such solid state spectra using the theory applicable to gas-phase samples. This approach is fundamentally incorrect, and frequently misleading, since the potential energy of a molecule in a solid must be perturbed by the presence of other molecules in the crystal lattice. Interpretation of solid state vibrational spectra by the chemist is therefore, more often than not, more a test of his intuition than of the data contained in the spectrum. This chemical intuition is dependent upon the amount of other analytical data available.

The increasing use of far infrared and laser Raman spectrometers has led to the detection of solid vibrations involving the movement of atoms and/or molecules as a whole. Such vibrations have no gas-phase analogy, and their existence emphasizes the need for a separate approach to solid state vibrational spectra.

A proper understanding of solid state vibrational spectroscopy requires a knowledge of the solid state physics of the process. The solid state physics of vibrational spectroscopy has been fully developed, though usually for simple single crystals of high symmetry. The chemist in contrast, often considers powders of complicated structure and low symmetry, which may therefore be too complex for their vibrational spectra to be explained by mathematical models. Nevertheless an understanding of simple models is a necessary first step in the interpretation of more complicated systems. This book attempts to explain the simple models and the way in which the theory can be extended to more complex models. It is hoped that it will provide an introduction to solid state vibrational spectroscopy for senior undergraduates and research workers unfamiliar with the field.

I would like to acknowledge the help of a number of my colleagues in Downing College and the University Chemical Laboratory. Above all I would like to thank Professor J. J. Turner, for his help, advice and encouragement throughout the preparation of this book. I am very grateful to Professor N. Sheppard, F.R.S. for reading the first draft of the book and making very helpful comments. Dr J. B. Pendry and

Preface

Dr J. K. Burdett read the greater part of the manuscript and made a number of very helpful criticisms. I would also like to thank the Salters' Company for their award of a Fellowship.

P.M.A.S.

Downing College
Cambridge
August 1971

1 Introduction

In this book the infrared and Raman spectroscopy of solids are discussed. Such spectra principally arise because of vibrational transitions in the solid, though transitions in the same region can arise due to solid state electronic and magnetic effects.

A number of reviews of the vibrational spectroscopy of solids are available [1–16] but often these reviews cover one aspect of the subject.

It is assumed that the reader of this book is familiar with the spectra of molecules in the gas-phase, and it is therefore useful in introducing the spectra of solids to consider how the spectrum changes when the molecules are in the solid-phase.

1.1 THE DIFFERENCES BETWEEN THE SPECTRA OF MOLECULES IN THE SOLID-PHASE AND MOLECULES IN THE GAS-PHASE

The molecules of a crystalline solid are different from the molecules in the gas-phase, in that they have a relatively fixed orientation with respect to the crystal axes, and it is this fixed orientation that is responsible for the appeciable differences between solid- and gas-phase spectra. Initially one may feel encouraged that by fixing the orientation of the molecule, the molecule is no longer able to rotate, and so one might expect the spectra of molecules in the crystal to be very sharp due to vibrational modes, no longer complicated by rotational fine structure. The study of any vibrational spectrum of a crystal soon disappoints us, since in general there is considerable broadening, with very different fine structure. Fortunately the vibrational frequencies of molecules in crystals are not very different from the corresponding frequencies in the gaseous state, but other differences, summarized below, are considerable:

(i) There is a great change in the shape of the vibrational bands, and often such bands have appreciably different intensities.

(ii) Gas-phase fundamental vibrations can be split into additional bands:

(a) *Static field splitting* (site group splitting) effects are a measure of the influence which the surrounding lattice, in its equilibrium configuration, exerts on the molecule. Non-degenerate internal vibrations can be

Introduction

shifted in frequency, and degenerate internal vibrations can also be *split*, because of the lower local symmetry of the crystal, which is no longer consistent with degeneracy. Inactive fundamentals may become active. The vibrations fall into two types: internal and external, where internal vibrations involve the stretching and bending, etc., of chemical bonds in the molecule (i.e. vibrations that are analagous with gas-phase vibrations), and external vibrations involve partial rotations and translations of molecules as a whole in the crystal lattice.

(*b*) *Correlation field splitting* effects are due to interactions with internal vibrations of *other* molecules in the same unit cell of the crystal. (*a*) plus (*b*) are sometimes called factor group splitting or unit cell group splitting. Both degenerate and non-degenerate internal vibrations may be split, since the potential energy will differ according as the internal vibrations are in phase, or partly out of phase in the unit cell. One fundamental vibration can be split into up to m bands where m is the number of molecules in the smallest volume, or primitive unit cell. Not all these m bands will always be infrared or Raman active, and some may be degenerate.

(*c*) *Internal–external vibration coupling*, involving thermally excited external vibrations, can cause non-degenerate internal vibrations to be *shifted*, and degenerate internal vibrations to be *split* (depending upon the symmetry of the internal and external vibrations). The importance of the effect increases as the vibrations of internal and external modes approach one another in value.

These splittings of the gas-phase fundamental vibration can be understood by considering the potential energy of the molecule in the crystal:

$$V = \sum_k V_k + \sum_k \sum_n V_{kn} + V_E + V_{Ek}, \quad (1.1)$$

where $\sum_k V_k$ represents the sum of the internal potential energies of the individual molecules in the crystal and may lead to static field splitting. The second term contains all the cross terms between the internal coordinates of different molecules and may lead to correlation field splitting. The last term contains cross terms between internal coordinates and external coordinates (given the symbol E) and may lead to internal–external vibration coupling.

(iii) A number of entirely new bands, not seen in the gas-phase spectra, will be observed in the low frequency region ($< 800\ \text{cm}^{-1}$), as a result of external vibrations (given by the term V_E in (1.1)). These external

Solid-phase spectra with gas-phase analogy

vibrations are a property mainly characteristic of the solid state, though they are sometimes observed in liquids [17–23], indicating a quasi-crystalline structure of the liquid.

(**iv**) The coupling between molecules in a crystal is often weak compared to the intramolecular forces, and a general crystalline field can then be introduced as a perturbation to the molecular field. The most important consequence of this is to introduce anharmonic coupling between external modes and internal modes, allowing combination bands of internal and external modes (which may be of a summation or difference nature).

(**v**) The number and vibrational activity of these combination bands may be very large indeed.

(**vi**) The vibrational frequencies of molecules in crystals are not very different from the corresponding frequencies in the gaseous state if the bonding is mainly of a van der Waals' type, but in molecules containing hydrogen atoms frequency changes may be large due to the formation of polymeric species through hydrogen bonding.

(**vii**) The size and nature of the crystal can have important effects on the observed spectrum.

(**viii**) The spectra of single crystals are more informative than those of powders, since complications due to powder scattering effects, and different crystalline orientations can be eliminated. Such single crystals are often, however, rather difficult to obtain.

(**ix**) Disordered and defect crystals can introduce additional fundamentals that can be very large in number. Local effects also occur due to specific defects.

(**x**) The applications of stress to a crystal may cause differences arising from:

(*a*) distortion of crystal symmetry, and consequent modification of selection rules,

(*b*) additional frequencies (in the same way as in (**ix**)),

(*c*) frequency shifts due to a change in the potential energy functions caused by pressure induced differences in interatomic distances, and

(*d*) changes in intensities of bands due to pressure induced dipole moment and polarizability differences.

(**xi**) Further splitting of gas-phase fundamental vibrations can result. Long range electrostatic forces in the crystal can lead to *longitudinal*-

Introduction

transverse splitting. Short range forces can lead to anisotropy of force constants causing different frequencies for different crystal directions. A variety of mixed modes, whose frequency depends upon the amount of mixing, may occur.

(**xii**) Near forward Raman studies may detect a variety of modes of mixed vibrational electromagnetic wave character, known as *polariton* modes, and these have a frequency different from those of the vibrational mode.

(**xiii**) A number of bands not due to vibrational transitions may be observed in the solid.

1.2 SOLID STATE VIBRATIONAL SPECTRA FOR WHICH THERE IS NO GAS-PHASE ANALOGY

There are a number of solid state vibrational spectra for which the model of a molecule perturbed by a crystal field cannot be applied. Such spectra include those of solids for which there are only external modes and no internal modes. Solids of this type may be covalent crystals such as diamond, or ionic solids such as sodium chloride.

1.3 THE WAVE MODEL FOR VIBRATIONAL EXCITATIONS IN SOLIDS

It can be seen that while the 'molecular' model can be applied with success to internal vibrations it cannot be applied to external vibrations, and thus the spectra of the type discussed in §1.2 cannot be analysed at all by this model. It is clear from the discussion so far that *all* vibrations are affected by the crystal environment. As a result of the periodic nature of the crystal lattice a range of possible vibrations occur where the atoms and/or molecules in adjacent unit cells move in phase or more or less out of phase. The energy difference between such vibrations would be expected to be small for internal vibrations which can be adequately described by a perturbation of a molecular system, because intramolecular forces are much stronger than intermolecular forces. Likewise the energy difference between such vibrations would be expected to be appreciable for external vibrations which completely depend upon intermolecular and/or interatomic or interionic forces. All crystal vibrations involve the entire lattice and are thus lattice vibrations (a term sometimes unfortunately only applied to external vibrations) and such vibrations

Wave model for vibrational excitations

can be considered as a wave propagating through the crystal lattice. The wavelength of this wave will take into account the phase difference between internal and external vibrations in adjacent unit cells. This wave model for crystal vibrations can be applied to *both* internal and external vibrations and will therefore be used throughout this book for the study of the vibrational spectroscopy of solids.

These waves that propagate through the crystal lattice may be represented by sine waves given by the equation

$$\phi = A \sin \frac{2\pi x}{\lambda}, \qquad (1.2)$$

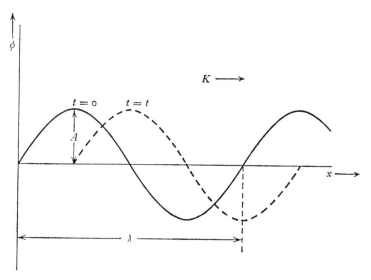

Fig. 1.1. Wave motion in a crystal lattice.

where the terms are explained in fig. 1.1. If the wave has a frequency ν, then in a time t the phase will have shifted by $\nu \lambda t$. Since $\phi = 0$, when $x = \nu \lambda t$, the wave can be represented in general by

$$\phi = A \sin \left[\frac{2\pi x}{\lambda} - X \right] \qquad (1.3)$$

and since $X = 2\pi \nu t$ when $x = \nu \lambda t$, (1.3) can be written:

$$\phi = A \sin \left[\frac{2\pi x}{\lambda} - 2\pi \nu t \right]. \qquad (1.4)$$

If a quantity K, called the *wave vector*, is defined and put equal to $2\pi/\lambda$

Introduction

and a quantity ω, called the *angular frequency*, is defined and put equal to $2\pi\nu$, then (1.4) becomes

$$\phi = A \sin(Kx - \omega t). \tag{1.5}$$

The wave vector clearly determines the direction of propagation of the wave. In this book, the wave vector of electromagnetic waves will be represented as k to distinguish it from K, the wave vector for wave motion in a crystal lattice due to vibrational, electronic, or magnetic excitations.

It is convenient to represent the wave function, ϕ, by a complex quantity

$$\phi = A e^{i(Kx - \omega t)}, \tag{1.6}$$

which may be written in terms of its real and imaginary parts:

$$\phi = A \cos(Kx - \omega t) + i A \sin(Kx - \omega t). \tag{1.7}$$

The real and imaginary parts can be added separately, and clearly represent the wave equation already described. Only the real part of the complex function is considered to have any physical meaning. To obtain the general equation of wave motion, the wave equation is differentiated:

$$\frac{\partial^2 \phi}{\partial x^2} = -K^2 \phi, \tag{1.8}$$

$$\frac{\partial^2 \phi}{\partial t^2} = -\omega^2 \phi, \tag{1.9}$$

which leads to the general equation

$$\frac{\partial^2 \phi}{\partial x^2} = \frac{K^2}{\omega^2} \frac{\partial^2 \phi}{\partial t^2}, \tag{1.10}$$

noting that $\omega^2/K^2 = v^2$ where v is the velocity of propagation of the wave.

This wave model for vibrational excitations in solids will now be used to examine such vibrations in detail. Of course not all these vibrations will be excited in infrared and Raman studies as fundamentals, but it will be seen later in the book that all these vibrations are of importance in second order effects.

2 Phonons and lattice vibrations

2.1 INTRODUCTION

In the first chapter the need to consider excitations in solids as propagating waves was explained. In this chapter the nature of these waves is examined in more detail. This examination involves consideration of:

(i) the interactions between atoms and molecules in the solid state (static mechanical properties);

(ii) the effect of the interactions in (i) on the vibrational modes of the solid (dynamical mechanical properties).

The vibrational modes of the solid can be excited by dynamic external forces, which can be mechanical or electromagnetic, and such excitation will be considered in detail later.

2.2 ELASTIC WAVES

The crystal may, to a first approximation, be considered as an elastic continuum [1, 24]. This involves treating the crystal as a homogeneous continuum with no particular reference to the atomic structure of the crystals. Thus the elastic continuum approximation will only be valid if the wavelength is much greater than a characteristic dimension of a unit cell in the crystal, which will mean that the wavelength will have to be greater than about 10^{-9} metres. The static mechanical properties can then be explained by macroscopic elastic constants or elastic moduli [2, 4].

Waves in the frequency region for which the continuum approximation is valid are identical with the longitudinal vibrations of an elastic string, and involve chains of atoms or molecules in the crystal acting as such a string. Such waves will be seen to be of only secondary importance in the vibrational spectroscopy of solids. The waves of primary importance are not those that involve atoms or molecules in the crystal acting as an elastic string but those wherein they are vibrating against each other, with the centre of mass of the cell fixed.

2.3 LATTICE DYNAMICS OF SIMPLE SYSTEMS

In this section the main features of the dynamical mechanical properties (lattice dynamics) of solids will be illustrated by means of simplified

Phonons and lattice vibrations

systems. Simple systems with one-dimensional lattices (linear chains or lines) and harmonic forces will be used, and the complications that occur in real systems will be discussed later. The simple systems to be discussed are composed of atoms, each atom being separated from each other atom by a distance a, known as the unit lattice vector, which moves one atom in the lattice into the next similar atom.

2.3.1 Elastic waves in a linear elastic band

It has been seen above that when the wavelength (λ) is of such a value that $\lambda \gg a$ the elastic continuum approximation is valid. An elastic band provides a good example of such a system. Fig. 2.1 shows a one-dimensional flat elastic band in an arbitrary state of strain [25], represented

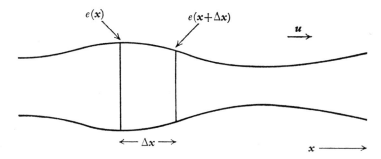

Fig. 2.1. One-dimensional elastic band.

by the varying thickness of the band. Consider an element of the band Δx, which has a displacement from its equilibrium position of $\boldsymbol{u}$.

$$\therefore \text{ Strain at one end} = e(\boldsymbol{x}) = \partial \boldsymbol{u}/\partial \boldsymbol{x}.$$

$$\text{Strain at other end} = e(\boldsymbol{x}+\Delta \boldsymbol{x}) = e(\boldsymbol{x}) + \frac{\partial e}{\partial \boldsymbol{x}} \cdot \Delta \boldsymbol{x},$$

$$= e(\boldsymbol{x}) + \frac{\partial^2 \boldsymbol{u}}{\partial \boldsymbol{x}^2} \cdot \Delta \boldsymbol{x}.$$

The forces at either end of the element oppose each other, and the resultant force is given by

$$F = c'(e(\boldsymbol{x}+\Delta \boldsymbol{x}) - e(\boldsymbol{x})) = \frac{c' \partial^2 \boldsymbol{u}}{\partial \boldsymbol{x}^2} \cdot \Delta \boldsymbol{x}, \tag{2.1}$$

where c' = longitudinal stiffness (force/strain).

Using force = mass × acceleration, and ρ = linear mass density of the band (dm/dx), then

$$F = \rho \Delta x \frac{\partial^2 u}{\partial t^2}. \qquad (2.2)$$

$$\therefore \frac{\partial^2 u}{\partial x^2} = \frac{1}{v_0^2} \cdot \frac{\partial^2 u}{\partial t^2}, \qquad (2.3)$$

where v_0 = velocity of sound = $(c'/\rho)^{\frac{1}{2}}$.

This equation can be seen to be identical to the general form of the wave equation (1.10) (where the velocity of sound v_0 replaces the velocity of light c). Solutions are therefore in the form of a travelling wave. It follows that

$$2\pi \nu = v_0 K. \qquad (2.3b)$$

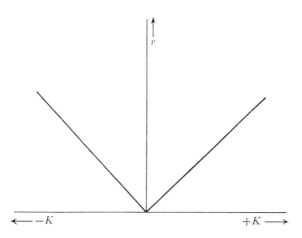

Fig. 2.2. Dispersion relation for vibrations of a one-dimensional elastic band.

This expression is known as the *dispersion relation* (the term comes from the dispersion of light) since it relates the frequency and wavelength. The relationship is clearly a straight line passing through the origin (fig. 2.2). The group velocity of a wave propagating through the system ($v_g = 2\pi \, \partial \nu / \partial K$) will be given by the slope of the dispersion curve, and in this case of the elastic band will clearly equal the phase velocity ($v_p = 2\pi \nu/K$).

2.3.2 Elastic waves in an infinite linear monatomic lattice

The discussion is now extended to include the short wavelength range where $\lambda \approx a$. Under these conditions the continuum approximation will

Phonons and lattice vibrations

not be valid and propagation at these short wavelengths will mean that the lattice structure of the crystal must be taken into account. The force that displaces lattice points from their equilibrium position must now take microscopic interactions into account in order to explain the short wavelength behaviour. When an atom is moved in such a displacement consideration should be made of movement of electrons as well as nuclei. The electrons are assumed to accommodate themselves so rapidly to the nuclear motion that the electronic state at a given moment is treated as if it were a unique function of the values of the nuclear coordinates at that moment. This neglection of the electronic coordinates is known as the *adiabatic approximation*. Born and Oppenheimer were the first authors [26] to examine this situation in detail, and further discussion can be found elsewhere [27, 28].

Using the adiabatic approximation, if m' is the mass of an atom in the lattice, the Hamiltonian for the whole lattice will be given by

$$\mathcal{H} = \tfrac{1}{2} \sum_{l} p_l^2/m' + V(x_1, x_2, ..., x_l), \qquad (2.4)$$

where the potential energy is a function of the position vectors x_l, and the kinetic energy is the sum of the kinetic energies of the individual atoms. The dynamical properties of the system can then be discussed using this Hamiltonian. For an infinite linear monatomic lattice, V will be given by

$$V(x_1, x_2, ..., x_l) = \sum_{l} f(x_{l+1} - x_l), \qquad (2.5)$$

which shows that V is a function only of the distance between the atoms at l and $l+1$. If the displacements of each of the atoms from their equilibrium positions is given by $u_1, u_2, ..., u_N$ (for N atoms), then the potential energy function can be expressed as a Taylor series in the displacements

$$\begin{aligned} V(u_1, u_2, ..., u_N) &= V_0 + \tfrac{1}{2} \sum_{ll'} u_l u_{l'} \frac{\partial^2 V}{\partial u_l \, \partial u_{l'}}, \\ &= V_0 + \tfrac{1}{2} \sum_{l} g(u_{l+1} - u_l)^2, \end{aligned} \qquad (2.6)$$

where g = second derivative of the interatomic potential function f, and the linear terms in the displacements vanish since an equilibrium state results when the atoms exactly occupy the lattice sites [29]. V_0 is taken to be zero. The potential energy above gives the potential energy for an infinite linear monatomic lattice of identical atoms of mass m joined by equal springs of force constant g. The expression (2.6) is only an

Lattice dynamics of simple systems

approximation to (2.5) since it neglects higher terms in the Taylor expansion, and this approximation is called the *harmonic approximation*, since the potential energy contains only harmonic terms [30].

The force that displaces lattice points from their equilibrium position is therefore given by

$$F = \sum_l g(u_{l+1} - u_l). \tag{2.7}$$

The situation is illustrated in fig. 2.3. Here the linear lattice motion of the atoms is restricted to a plane and transverse motion considered. As in (2.2)

$$F = m' \cdot \frac{\partial^2 u}{\partial t^2}, \tag{2.8}$$

$$\therefore \frac{\partial^2 u}{\partial x^2} = \frac{1}{m'} \sum_l g(u_{l+1} - u_l). \tag{2.9}$$

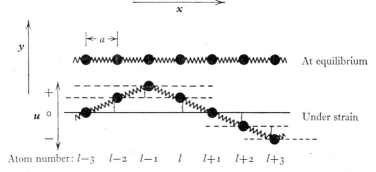

Fig. 2.3. Transverse elastic waves in an infinite linear monatomic lattice. (Longitudinal elastic waves could occur in the *x* direction.)

Since in the harmonic approximation the force effectively extends only to nearest neighbour atoms (hence the values of l and $l+1$ above), solutions for (2.9) can be found in the form of a travelling wave

$$u_l = u_{l_0} e^{i(Kx - 2\pi \nu t)}. \tag{2.10}$$

The linear monatomic lattice is defined by

$$x = la \quad (l = 1, 2, ..., N) \quad \text{(for } N \text{ atoms)}, \tag{2.11}$$

$u_{l_0} = u_{(l+1)_0}$ is taken to be the amplitude of the travelling wave and will be denoted u_0. Thus

$$u_{l+1} = u_0 e^{i(K(l+1)a - 2\pi \nu t)}, \tag{2.12}$$

where the vector $\mathbf{a}$ has magnitude a.

Phonons and lattice vibrations

Substituting values of u_{l+1} and u_l of the form of equation (2.12) into (2.9), and solving for ν there results [24, 25, 31, 32]

$$2\pi\nu = \frac{(4g)^{\frac{1}{2}}}{m'} |\sin \tfrac{1}{2} Ka|, \qquad (2.13)$$

where g is related to the longitudinal stiffness of (2.1) by $c' = ga$.

The dispersion relation can now be computed from (2.13) and is shown in fig. 2.4. As in fig. 2.2 both positive and negative values of K are shown to allow propagation to the right or to the left. The difference between the lattice and a continuum can be clearly seen from this relation. In

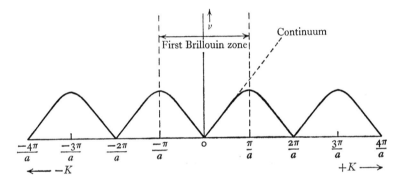

Fig. 2.4. Dispersion relation for vibrations of an infinite one-dimensional monatomic lattice.

the case of a lattice, ω can be seen to be a periodic function of K, unlike a continuum where ω increases linearly with K without limit. For a lattice the maximum frequency is reached when $K = \pi/a$. The reason for the levelling of the dispersion relation towards the maximum frequency can be understood when the microscopic wave motion is considered. When K is small (λ large), the wave motion is similar to that of a linear elastic band (fig. 2.5(a)), and the continuum approximation holds. When K becomes large however, account must be taken of the lattice structure, and in the limit the motion involves the atoms vibrating back and forth against each other, and a standing not a travelling wave results (fig. 2.5(b)). The situation is similar to Bragg reflection of X-rays and the Bragg condition is satisfied [24]. There is a cut-off in the possible values of K which are restricted to the range $-\frac{\pi}{a} < K \leqslant \frac{\pi}{a}$ known as the first Brillouin zone. This restriction results because an elastic wave has a meaning for only the atoms themselves, and since wave motion implies

Lattice dynamics of simple systems

crests of motion moving down a line at a certain velocity, wavelengths shorter than $\lambda = 2a$ cannot propagate. A wave motion with wavelength shorter than $\lambda = 2a$ is not meaningless however, but is entirely equivalent to a wave of wavelength greater than $\lambda = 2a$ (fig. 2.5(c)).

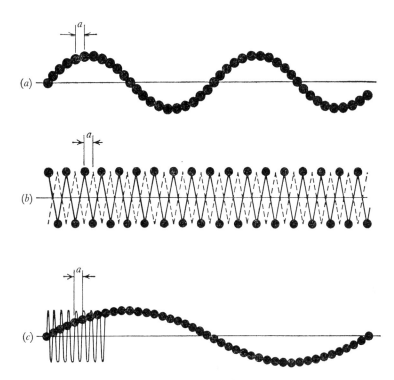

Fig. 2.5. Wave motion in an infinite linear monatomic lattice.

Since elastic waves in an infinite linear monatomic lattice are for most K values identical to the elastic waves in a linear elastic band they form typical sound waves, and are thus known as *acoustic waves*. The $K = 0$ acoustic wave corresponds to a translation of the entire lattice.

2.3.3 Elastic waves in an infinite linear diatomic lattice

With two or more kinds of atom per primitive cell the dispersion relation shows new features. Consider a unit cell consisting of two atoms of mass m_1 and m_2 located at even and odd numbered lattice points, the atoms being separated by the distance a (fig. 2.6). The equations of motion [24, 25, 31, 32] can be calculated in the same way as has been done for

Phonons and lattice vibrations

the monatomic lattice using the harmonic approximation. Such a calculation leads to the result

$$4\pi^2\nu^2 = g\left(\frac{1}{m_1}+\frac{1}{m_2}\right) \pm g\left[\left(\frac{1}{m_1}+\frac{1}{m_2}\right)^2 - \frac{4\sin^2 Ka}{m_1 m_2}\right]^{\frac{1}{2}}. \quad (2.14)$$

Two solutions are clearly possible, known as the *acoustic* (the same as the previous case) and the *optical* branches. The optical branches are so named because they are the waves that give rise to fundamentals in infrared and Raman studies. Fig. 2.8(a) illustrates a typical wave of the acoustic

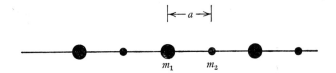

Fig. 2.6. An infinite linear diatomic lattice.

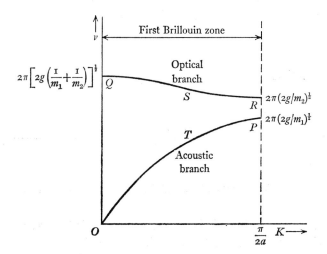

Fig. 2.7. Dispersion relation for vibrations of an infinite one-dimensional diatomic lattice. (Limiting frequencies shown for $m_1 > m_2$.)

branch at T (fig. 2.7), and fig. 2.8(b) illustrates a typical wave of the optical branch at S (fig. 2.7). As before the waves of the acoustic branch are seen to be identical to those of a linear elastic band, except when K becomes large and a standing wave results at the maximum value of K (P in fig. 2.7) which involves the heavy atoms vibrating back and forth against each other with the light atoms fixed (fig. 2.8(d)). In the case of

Lattice dynamics of simple systems

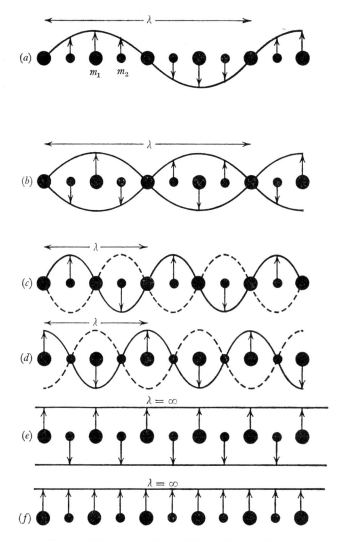

Fig. 2.8. Wave motion in an infinite diatomic lattice.

a diatomic lattice the maximum value of K will be $\pi/2a$, which defines the first Brillouin zone for waves propagating in one direction. The waves of the optical branch, however, can be seen to be quite unlike those of the acoustic branch [34] and involve, *for all values of K*, the atoms vibrating back and forth against each other. When $K = \pi/2a$ (R in fig. 2.7) the light atoms vibrate back and forth against each other with

Phonons and lattice vibrations

the heavy atoms fixed (fig. 2.8(c)). Clearly if $m_1 = m_2$, P and R become coincident, and if m_1 tends to infinity the acoustic branch disappears ($P = (2g/m_1)^{\frac{1}{2}} = 0$), and the optical branch becomes flat (QR is a straight line) since in the harmonic approximation next nearest neighbour interactions are zero. When $K = 0$ the optical branch (Q in fig. 2.7) involves

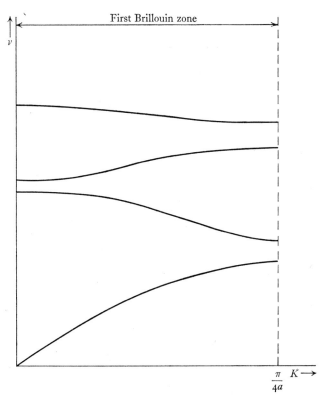

Fig. 2.9. Dispersion relation for vibrations of an infinite one-dimensional tetraatomic lattice.

the two types of atom vibrating rigidly against each other (fig. 2.8(e)), and the acoustic branch (O in fig. 2.7) involves a translation of the entire lattice (fig. 2.8(f)).

2.3.4 Elastic waves in an infinite linear multiatomic lattice

Consider a unit cell consisting of n atoms of mass m_1, m_2 to m_n. The equations of motion can be calculated [31] as before and when solved give one acoustic branch and $(n-1)$ optical branches. Fig. 2.9 shows the

Lattice dynamics of simple systems

dispersion relation for a unit cell containing 4 atoms. This dispersion relation can also be expressed by means of an extended Brillouin zone diagram (fig. 2.10). This diagram shows values of ν outside the first Brillouin zone. It was however pointed out when considering an infinite linear monatomic lattice that a wavelength outside the first Brillouin zone was not meaningless but was entirely equivalent to a wave in the first Brillouin zone. Thus fig. 2.10 can be converted into the reduced Brillouin

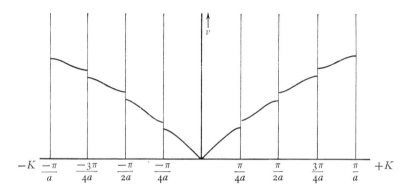

Fig. 2.10. Dispersion relation for vibration of an infinite one-dimensional tetraatomic lattice shown on an extended Brillouin zone diagram.

zone diagram of fig. 2.9 by drawing the part between $\pi/4a$ and $\pi/2a$ (fig. 2.10) as the part between $\pi/4a$ and 0 (fig. 2.9) corresponding to the lowest frequency optical mode and so on. The reduced Brillouin zone diagram of fig. 2.7 could be converted into an extended Brillouin zone diagram in the same way. In the new extended diagram for fig. 2.7 there will be a frequency gap (energy gap) between P and R (unless $m_1 = m_2$) depending upon whether the heavy (P) or the light (R) atoms vibrate. S (fig. 2.8(b)) could be considered to have $\lambda = 4a$, and Q (fig. 2.8(e)), $\lambda = 2a(K = \pi/a)$ in the section of the extended diagram between $\pi/2a$ and π/a.

2.4 THE FINITE LATTICE AND BOUNDARY CONDITIONS

When a linear lattice is considered K cannot take all the values given by the dispersion relation. This follows from the earlier discussion about the significance of the Brillouin zone. In such a case the solution may be

Phonons and lattice vibrations

obtained from that of an infinite crystal by applying appropriate boundary conditions. This solution will generally be determined by the conditions at the extremities of the lattice.

2.4.1 Fixed boundary

Consider a linear monatomic lattice consisting of $N+2$ atoms, the end atoms (labelled 0 and $N+1$) being *fixed*. The length of the lattice will be $(N+1)a$, and $u_0 = 0$ and $u_{N+1} = 0$ $(x = la)$.

The solution may be written as

$$u = u_0 e^{iKx} = u_0(\cos Kx + i \sin Kx), \qquad (2.15)$$

and satisfying the conditions above, $u = 0$ when $x = 0$, and $u = 0$ when $x = (N+1)a$ if $K(N+1)a = n\pi$ (where n is an integer);

i.e.
$$K = \frac{n\pi}{(N+1)a} = \frac{n\pi}{L}, \qquad (2.16)$$

where L = length of the lattice. For the first Brillouin zone n has the values $n = 1, 2, 3, ..., N$.

(2.16) therefore shows that in order to have both ends fixed, there must be an integral number of half wavelengths in the length of the crystal. If $N = 3$ for example [31], only three values of K will be allowed in the first Brillouin zone.

2.4.2 Cyclic boundary

All the solutions obtained for the case of a fixed boundary above will consist of standing waves, in contrast to the infinite lattice solution when both travelling and standing waves were obtained. By considering cyclic boundaries, it is possible to obtain solutions for the finite lattice that include both standing and travelling waves.

If atoms 0 and $N+1$ are no longer fixed, but joined so that the line of $N+2$ atoms forms a circle or other closed curve (fig. 2.11), cyclic or periodic boundary conditions result. The situation now corresponds to an infinite lattice that repeats itself in groups of $N+2$ points, the length of each group being $(N+2)a$. The only requirement now is that

$$u_l = u_{l+N+2}. \qquad (2.17)$$

The solution may be written as

$$u_l = u_0 e^{iKx}, \qquad (2.18)$$

Finite lattice and boundary conditions

and therefore
$$u_{l+N+2} = u_0 e^{iK[x+(N+2)a]}. \tag{2.19}$$

Satisfying the conditions in (2.17) above, and since $e^{iK(N+2)a} = 1$, then

$$K = \frac{2n\pi}{(N+2)a}, \tag{2.20}$$

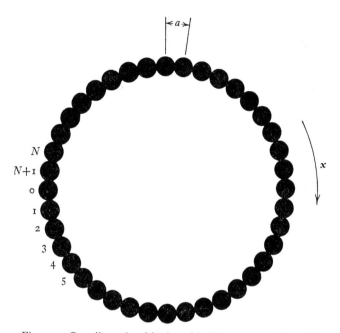

Fig. 2.11. One-dimensional lattice with $N+2$ repeat units, with cyclic boundary conditions.

where n is an integer. The solution is now a travelling wave. For the first Brillouin zone n has the values

$$n = \mp 1, \mp 2, \mp 3, ..., N+2.$$

In this case negative values of n are considered since in the case of a travelling wave the wave can propagate either to the left or to the right. Thus while the allowed values of K are twice as far apart in K space, the first Brillouin zone is twice as big, so the same number of modes are obtained for both cyclic and fixed boundary conditions. The modes obtained for both boundary conditions and for monatomic and diatomic lattices are shown in fig. 2.12.

Phonons and lattice vibrations

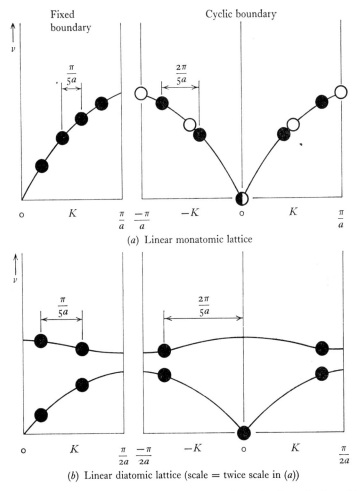

Fig. 2.12. Comparison of fixed versus periodic boundaries for linear lattices of length $5a$ (6 atoms for fixed, and 5 for cyclic boundaries). Case of cyclic boundary for 4 atoms for monatomic lattice shown (O) gives 5 modes, but $K = \pi/a$ and $-\pi/a$ modes are identical.

2.5 THE THREE-DIMENSIONAL LATTICE, LONGITUDINAL AND TRANSVERSE WAVES

The theory is now generalized to describe the vibrational modes of a general lattice in three dimensions. These extra dimensions complicate the problem algebraically. For the linear lattice, motion of the atoms was restricted to a plane and only transverse motion was considered. In

Three-dimensional lattice

three dimensions, transverse motion will be possible in two dimensions, and in addition longitudinal displacements of the atoms must also be considered. It can be shown that the transverse motion in the two dimensions (two planes at right angles to one another) will be mutually independent and the longitudinal displacements (for small displacements) will be independent of the transverse motion. Theoretical studies [35] show, however, that in general waves in lattices are neither transverse nor longitudinal, especially when K is very small, and the nature of waves is determined by the details of the interatomic forces and/or intermolecular forces [36]. A more detailed discussion of such general waves will be left until chapter 4. In certain special directions the normal vibrations will be strictly transverse and/or longitudinal because of symmetry, since the vibrations involve restriction of the motion of the particles to a plane. For a face-centred cubic crystal the (100), (110) and (111) planes correspond to the longitudinal and two transverse modes, and the dispersion relations calculated as three separate linear lattice problems. The two transverse optical modes will be degenerate, but the longitudinal optical mode will occur at a higher frequency at $K = 0$ because such vibrations are associated with a finite macroscopic electrostatic field (to be explained fully in chapter 4). The difference in frequency between the transverse and longitudinal modes is the *longitudinal-transverse splitting* described in chapter 1. Fig. 2.13 illustrates the dispersion relation for a typical three-dimensional case.

The solution of the three-dimensional lattice problem with cyclic boundaries [30] was first carried out by Born and von Kármán [37]. These conditions involve bending the lattice in one dimension to form a cylinder, which can be joined up into a torus. The third dimension cannot be realized physically since the torus cannot be deformed further [29], and so Born and von Kármán conditions cannot be satisfied simultaneously, for topological reasons. They are, however, the only mathematical device for simplifying the problem, and it can be shown [38] that this causes negligible error in a sufficiently large crystal.

For a three-dimensional crystal containing N atoms with n atoms per unit cell there will be $3N$ normal modes of vibration distributed on $3n$ branches, $3n-3$ of which will be optical modes and 3 acoustic modes. This results from multiplying the modes for the linear multiatomic lattice by three.

Phonons and lattice vibrations

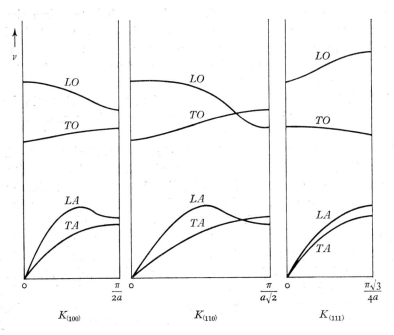

Fig. 2.13. Dispersion relation for vibrations of an infinite three-dimensional diatomic lattice for a typical face-centred cubic crytsal. The three diagrams show dispersion relations for phonons in the (100) (110) and (111) directions. K is calculated from (3.15) where $|t_1| = |t_2| = |t_3| = 2a$ and, at the zone boundary, $x_1 = N_1$, $x_2 = N_2$ and $x_3 = N_3$. L and T refer to longitudinal and transverse waves, and O and A to optical and acoustic branches.

2.6 EXTERNAL AND INTERNAL VIBRATIONS

So far only the vibrations of atoms has been considered. When the solid contains molecules two possibilities result.

(i) The molecules vibrate as a whole, the molecular groups performing rigid body motion. Such vibrations are known as *external vibrations*, or *lattice vibrations*, and are separated into two types. *Translational modes* involve translation of the molecular groups, and correspond to the vibrations of atoms discussed above. *Rotational* or *librational* modes involve quasi-rotation of the molecular groups about their centres of gravity and therefore require the presence of a polyatomic group in the crystal. Such a rotational mode could clearly never be considered to approximate to the motion of a linear elastic band and so could not

External and internal vibrations

contribute an additional acoustic mode. Rotational modes are optical modes, and fig. 2.14(a) illustrates a typical example.

The distinction between the two types of external vibration is not as clear as might at first be imagined. For a centrosymmetric molecular site (fig. 2.14(b)), a rotational displacement keeps the centre of mass of the molecule fixed, and preserves the centre of inversion. A translational displacement for such a case would displace the centre of mass and destroy the centre of inversion. This differentiation between the two

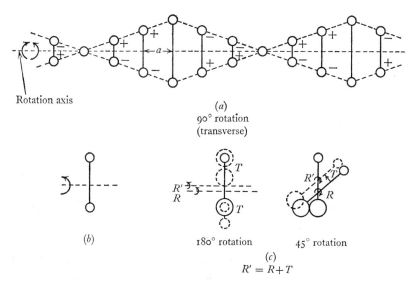

Fig. 2.14. Rotational modes for a diatomic molecule in a crystal lattice. (In (a) and (b) the atoms are the same, in (c) they are of different mass.)

types of mode allows them to be accurately separable for $K = 0$ (this is not the case when $K \neq 0$). When the site is no longer centrosymmetric (fig. 2.14(c)), the centre of inversion is lost and rotation about the centre of mass (R) can be related to rotation about the centre of distance (R') and a translation (T). In such a case the rotational mode can couple with a translational mode and a mixed mode result. These mixed modes will be discussed in more detail later.

(ii) The atoms that comprise a molecular group vibrate within the molecular group. Such vibrations involving non-rigid molecular motions and allowing for coupling between different groups and crystal periodicity are known as *internal vibrations*. It is the coupling between the different

Phonons and lattice vibrations

groups which causes the difference between these vibrations and gas-phase vibrations, and the crystal periodicity will determine the dispersion relation for the modes. These modes can only be optical ones, and their presence adds additional optical modes to the dispersion relation. Fig. 2.15(a) illustrates a typical optical mode resulting from an internal symmetric stretching vibration of a diatomic molecule in a crystal, and fig. 2.15(b) illustrates a typical optical mode resulting from an internal symmetric deformation mode of a triatomic molecule in a crystal.

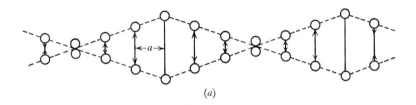

(a)

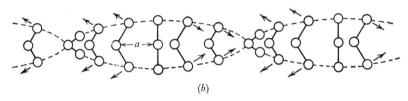

(b)

Fig. 2.15. Internal optical modes. (a) Symmetric stretching vibration of a diatomic molecule. (b) Symmetric deformation vibration of a triatomic molecule. The modes are drawn as transverse for convenience.

For a three-dimensional crystal containing N atoms with n atoms contained in m molecules per unit cell there will be $3N$ normal modes of vibration distributed on $3n$ branches, $3n-3$ of which will be optical modes and 3 acoustic modes. Of the optical modes $(3n-6m)$ will be internal modes, $3m$ rotational modes, and $3(m-1)$ translational modes. A typical dispersion relation for such a case is shown in fig. 2.16.

In an infinite crystal it is impossible to have rotations of the entire crystal, and while for a finite crystal three such rotations are possible. These rotations are not considered because a finite crystal with Born and von Kármán cyclic boundaries behaves as an infinite crystal. This is similar to the case of a diatomic molecule of two point masses, for which one does not consider rotations about the internuclear axis.

External and internal vibrations

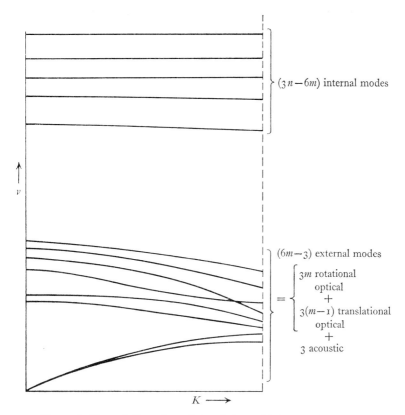

Fig. 2.16. General dispersion relation (schematic) for a molecular crystal containing n atoms in m molecules per unit cell.

2.7 THE QUANTIZATION OF LATTICE VIBRATIONS

Early studies of lattice vibrations were concerned with their ability to transport energy in the form of heat. The presence of such waves caused a contribution to the specific heat of the solid concerned. Quantum theory requires this energy to be quantized by saying that the energy of a lattice vibration must be a multiple of $h\nu$. Such quanta are known as *phonons* [39], by analogy with the quantum of electromagnetic radiation, the photon. Phonon waves and lattice waves are thus not quite synonymous terms, since phonon waves have a more precise connotation than lattice waves in that they involve standing waves quantized in the various modes.

The energy of the phonon is sometimes found stated in the literature [24, 2] as $\hbar K v_0$, where v_0 is the velocity of sound in the lattice. This

25

Phonons and lattice vibrations

expression is only strictly correct for elastic waves in an elastic band (§2.3.1), though approximately correct for acoustic waves (when the wavelength is long). The expression does not, however, apply to optical waves since the dispersion relation clearly shows that the velocity must vary with K. It will be seen later that the velocity of optical waves lies between the velocity of light in one extreme, and the velocity of sound in the other. The energy of a phonon is always given by $h\nu$.

Phonons from two sources will be important.

(i) Phonons created by external excitation (e.g. infrared, and Raman spectroscopy). Only optical phonons can be created by external excitation in first order processes.

(ii) Phonons created or destroyed by raising or lowering the temperature, i.e. thermal phonons. Both optical and acoustic phonons can be produced by thermal effects. Sound waves are acoustic phonons, and they can contribute significantly to warming a crystal only at very low temperatures or high intensities. Thermal phonons are important in second order spectroscopic processes (difference bands) and for their effect on normally elastic scattering processes (e.g. X-ray and neutron scattering).

The momentum of the phonon calculated from the de Broglie relation is $\hbar K$. In fact a phonon does not really have momentum, except in the case of an acoustic phonon with $K = 0$ (which is a translation of the entire crystal), and so the physically more exact term 'conservation of wave vector' rather than 'conservation of momentum' is used.

In general for any process involving phonons two rules apply:

(i) conservation of energy, and
(ii) conservation of wave vector.

Information about phonon waves can be obtained experimentally from:

(i) specific heat measurements,
(ii) compressibility measurements,
(iii) vibrational spectroscopic studies,
(iv) inelastic scattering of neutrons,

and these experimental results can be compared with dispersion relations calculated theoretically. The theoretical calculation of a dispersion relation in real crystals requires a knowledge of the interatomic and intermolecular forces and the ways in which this problem can be tackled will now be discussed.

2.8 LATTICE DYNAMICS OF REAL CRYSTALS

The analysis of the vibrations of atoms and molecules about their positions of equilibrium in *real* crystals is a very difficult problem. Approximate numerical solutions have been obtained for only the simplest lattices and these solutions have involved making very drastic assumptions regarding the interatomic and intermolecular forces.

The systems that have been studied fall into two types.

(i) Crystals with relatively simple structures and a small number of atoms in the primitive unit cell.

(ii) Crystals with complex molecules and ions, often with complicated structures and a large number of atoms in the primitive unit cell.

Systems of type (i) have been of interest in the study of the basic physics of the interatomic forces, while those of type (ii) have been of interest to molecular physicists and spectroscopists. While systems of type (ii) have been considered in terms of an individual molecule interacting with its environment, the lattice dynamics of all crystals can be considered in terms of cooperative vibrations of all atoms in the crystal using the ideas of external and internal modes introduced in §2.6.

2.8.1 *Models for the potential energy of a crystal*

The early work on lattice dynamics has been reviewed by Debye [40] and Born [41]. The phenomenological framework developed by Born and Huang [34], which considers cooperative vibrations of all atoms in the crystal, has been extensively used in the analysis of dispersion relations particularly for simple crystals. The application of this theory has been extensively reviewed [34, 42–7]. In this approach the potential energy of the crystal is expanded in powers of the displacements of the atoms from their equilibrium positions (see (2.6)), and only harmonic terms considered.

The concept of force constants that couple one atom or group to another and are an integral part of the harmonic term (see (2.6)) is central to the model of Born and Huang. The main difficulty with the use of the Born model for predicting experimental results is the need for a detailed knowledge of the interatomic and intermolecular forces in order to calculate these force constants. Anharmonic effects have been calculated by taking into account appropriate higher order terms (see (2.6)) [49–55].

Phonons and lattice vibrations

In recent years a microscopic approach starting from the basic quantum theory of solids has been introduced for the study of lattice dynamics. The approach was first applied to the study of simple metals [56–77] by replacing the ionic potential by a pseudopotential or model potential [71, 77], and using perturbation theory, taking many electron effects into account by performing a self-consistent screening calculation. There is no fundamental justification [78] for treating metals by a microscopic theory and insulators and semiconductors by a phenomenological theory, and a quantum theoretical approach for solids in general has been introduced [79–81]. In this model the phenomenological model is reformulated in terms of potential energy functions [82] leading to nuclear–nuclear interactions for all orders being considered, and the concept of force constants is not used at all. It has been shown [83] that a modified phenomenological model involving more detailed potential functions can correctly reproduce all the features of the quantum theoretical results.

2.8.2 Interatomic and intermolecular forces

Relatively little progress has been made in quantitative calculations based on an understanding of the interatomic and intermolecular forces. These forces are often represented by arbitrary parameters such as coefficients of the potential energy function, and the values of these parameters 'fitted' to agree with experimental results. Thus when the classical theory was used and the results compared with experiment it was always found that much longer range forces were required to fit the results, and the parameters were suitably 'fitted' to the results.

More work has been done on crystals with ionic forces than on crystals with covalent forces. For ionic crystals in the simplest microscopic model the ions were treated as unpolarizable, undistortable, point charges attracted by longer range coulomb attractions and held apart by shorter range overlap forces. A lattice vibration was considered in terms of a uniform distortion of the crystal in which the positive ions move against a rigid sub-lattice of negative ions. Long wavelength ($K \to 0$) lattice vibrations would therefore depend upon the longer range Coulomb attractions, and shorter wavelength ($K \to$ Brillouin zone limit) lattice vibrations would depend upon the shorter range overlap forces. To take the much longer range forces discussed above into account this classical model of Born and Huang was modified by the addition of other forces depending upon the system studied. The original rigid ion approach (used by Kellermann [84] in his studies of the alkali halides) was modified by the shell models [85–95] which, in various different ways, take into account the long range forces that result from the polarizability of the ions in the lattice. This polarization occurs when the ions are displaced in a lattice vibration. The ion displacement u causes each ion to acquire a dipole moment $\partial \mu_d$ as a result of distortion from overlap with nearest neighbour ions and distortion from Coulomb interactions with other ions [96]. The charge on each ion, q, is therefore caused to deviate appreciably from a multiple of electronic charge (this can reduce the apparent charge in alkali halides up to 30 per cent) to a value q^* given by

$$q^* = q - \partial \mu_d / u. \qquad (2.21)$$

The use of these shell models has been reviewed [97, 33], and so far the extended-shell model has been the most successful and widely used [98]. The quantum mechanical justification of these models has been discussed [99–103], and the model derived from a microscopic basis [104].

For covalent crystals the forces are much less well defined. Phillips [105–7] has discussed the covalent bond in elemental and partially ionic crystals. For a typically covalent crystal q (see (2.21)) is zero, but in any heteroatomic system electronegativity differences would be expected to give a q^* that was finite. q^* has been found to be as large as $0.58e$ in some covalent compounds [3]. For the internal modes of covalent compounds strong intramolecular covalent forces will be involved, but the K dependence of such modes and the external modes will depend upon the orientation and inductive forces that give rise to a finite q^*. Thus the general results of a model based on Coulomb forces would be expected to be relevant to a covalent crystal with a finite q^*. While the shell model has been applied to covalent crystals with some success it is not entirely satisfactory [107, 89]. For elemental covalent crystals there will be no electronegativity difference and thus no orientation or inductive forces giving rise to a finite q^*.

The difficulty with the more complex phenomenological models discussed above for ionic and covalent crystals discussed above is that the experimental data cannot be explained solely in terms of electrostatic forces acting on point charges and point dipoles, and a considerable number of extra forces of unknown origin must still be postulated to explain the data. The electron motion, which was taken into account above by considering a polarizable ion, has been considered [108, 109] in terms of a microscopic response function. An important difference between the forces in metals and those in insulators arises because the

Phonons and lattice vibrations

perfect screening of an ion by conduction electrons in a metal causes the forces between the ions to be short ranged, while for insulators long range forces have been seen to play a fundamental role. In the case of covalent crystals the electron motion has been taken into account by considering a covalent bond–charge model [105, 106].

The number of general force constants has been reduced by assuming special forces between atoms or groups. These special forces, in order of increasing complexity, are known as central forces (two body forces), angular forces (three body forces), and torsional forces (four body forces).

Classical phenomenological models have been used to calculate the lattice dynamics of metals [110–13], anisotropic crystals [114], alkali halides, alkaline earth halides [115], mercurous halides [116], semi-ionic compounds [117], molecular solids [118–28], and complex crystals [47, 45, 129–32], and in all cases the lattice dynamical calculations are complex, needing appreciable computation [133] (even for simple atomic cubic lattices [31] the calculation is quite tedious).

The lattice dynamics of the lightest of molecules causes difficulties when the classical approach is used. The lattice dynamics of solid hydrogen or deuterium [134, 135] cannot be treated by the harmonic approximation because the molecules are so light and the intermolecular forces so weak that such a calculation gives imaginary energies of excitation. In such systems the dominant interactions between molecules in the solid are those between the electric quadrupole moments of the molecules [136], these interactions being much larger than valence and van der Waal's interactions.

2.8.3 The effect of finite crystal size

When a finite crystal size is considered the influence of the crystal surfaces must be taken into account. The crystal surfaces affect the lattice dynamics of the crystal because the translational symmetry of the crystal at any direction perpendicular to the surface is lost. This causes an asymmetric environment for atoms and/or molecules near the surface which are thus subjected to different forces from those in the bulk. In fact the long range forces discussed above cause atoms and/or molecules quite far below the surface to be affected by the surface, causing the modes of the crystal to be dependent upon the size and shape of the crystal.

The introduction of a crystal surface into the problem presents the possibility of a new type of lattice vibration, namely the *surface mode*. Such a mode was first studied by Lord Rayleigh [137] who considered

Lattice dynamics of real crystals

waves on the stress-free surface of a semi-infinite isotropic medium. These acoustic waves, known as *Rayleigh modes*, have displacement components that exponentially decrease perpendicular to the surface. Love [138] discovered another type of acoustic surface wave involving a transverse shear deformation in an isotropic infinitely long and thick slab resting on a different semi-infinite isotropic medium. Such acoustic surface modes, which cannot be excited by infrared absorption (though they can be optically excited by surface heating from laser illumination [139–41]), have been of interest because of their potential use in u.h.f., v.h.f., and microwave devices [142–4]. Special acoustic modes known as gap modes can be found near the two-dimensional Brillouin zone edge in monatomic crystals, which have frequencies either below those of all the bulk modes or fill a gap between bands of different types of bulk mode [145]. The mixing of surface and bulk modes has been discussed [146].

For monatomic crystals only the acoustic surface modes discussed above can exist, but for crystals with two or more atoms and/or molecules in the unit cell two types of surface mode have been considered:

(i) acoustic surface modes for the lattice treated as a continuum, and
(ii) optical surface modes.

The optical surface modes [147], which have no analogy in continuum theories, can be excited by infrared absorption. These optical surface modes can be considered as localized vibrations at the surface (with an exponentially decreasing amplitude perpendicular to the surface) and can be described by a wave vector K_x parallel to the surface. The atomic and/or molecular displacements of the optical surface modes can be considered parallel and perpendicular to the surface.

Lattice dynamical models for the calculation of surface mode frequencies are more difficult to carry out than calculations based on continuum models. Calculations have involved monatomic simple cubic lattices [148, 149], and diatomic cubic lattices [150–2]. In diatomic cubic lattices, atomic displacements of the optical surface modes have been calculated parallel and perpendicular to the surface [150], and the frequencies of these modes calculated for rocksalt of 15 atomic layers thickness [151] and semi-infinite thickness [152]. The attenuation depth of the optical surface modes in rocksalt has been found to extend to 50 atomic layers [152], with frequencies very close to the bulk modes for the semi-infinite thickness of crystal, but with frequencies different to

Phonons and lattice vibrations

the bulk modes (because of interaction between the two surfaces) for a crystal only 15 atomic layers thick [151].

Additional surface modes that arise from adsorbed atoms or from a macroscopic layer of one material lying on a substrate of another material [153] have been discussed.

The surfaces of a crystal can be regarded as extended defects in the crystal [154], and in fact the surface modes discussed above can interact with localized modes caused by defects [147, 155].

2.8.4 The effect of crystal disorder and defects

Crystal samples frequently contain impurities, disorder, distortion [156], and stacking faults [157], and these can cause striking macroscopic effects. These faults, some of which can be observed directly [158, 159], destroy the translational symmetry of the crystal hence changing the normal modes from their usual plane wave form and breaking down the conservation of wave vector rule (the important effect that this has on the spectroscopic properties will be discussed later).

The general mathematical theory of the effect of faults on lattice vibrations has been investigated in some detail, and a number of reviews are available [2, 160–8]. Most studies have involved the use of a simple model using a linear-chain lattice with nearest neighbour harmonic forces. As expected from the previous study of perfect lattices, the two- and three-dimensional cases will be much more difficult to study and therefore these systems have received little attention [169].

Various specific types of lattice have been discussed. Metallic crystals have been considered by using a model consisting of a monatomic linear chain containing an atomic impurity [170], ionic crystals by a model consisting of a diatomic linear chain containing an atomic impurity [171, 172], and valence crystals by a model consisting of a monatomic linear chain containing a molecular impurity [173]. Three models have been used to study molecular crystals, one model involves an atomic impurity [174], the second involves a molecular impurity [175], and the third a model involving a crystal composed of homonuclear diatomic molecules, one of which has a different force constant from the others [176]. The third model has the novel feature of being applicable to pure molecular crystals when the molecule with the different force constant is a molecule of the crystal in an excited electronic state or a molecular ion.

Atoms associated with a point defect have also been treated by partitioning the defect region and treating it as a pseudo-molecule coupled

Lattice dynamics of real crystals

to the rest of the crystal [177–83]. The effect of the motion of interstitials, vacancies and dislocations has also been considered [184, 185].

The presence of a point defect in an otherwise perfect crystal can modify the modes in two possible ways.

(i) The modes of the perfect crystal will be modified, but the modified modes lie within the bands of perfect lattice frequencies, the motion being transmitted throughout the crystal. The modes of the perfect lattice will be most strongly perturbed near the defect. Such modified modes are known as *band modes*, or *non-localized modes*.

(ii) The modified modes lie outside the bands of perfect lattice frequencies. Such modes are known as *localized modes*, since they cannot be transmitted throughout the crystal and are highly localized around the defect. These localized modes are sometimes classified according as they occur at higher frequencies than the perfect lattice frequency (localized modes), or between bands of allowed perfect lattice frequencies (localized gap modes, or gap modes).

The possible modified modes depend upon the mass of the defect atom. For a monatomic crystal if the defect atom is heavier than the atom it replaces, a non-localized mode results (fig. 2.17(a)), and if it is lighter, a localized mode results (fig. 2.17(b)), as long as the force constants between the defect atom and its neighbours are similar to those between pairs of host crystal atoms. If the force constants between the defect atom and its neighbours are weak a resonant behaviour [164, 186–8] can occur if the defect mode lies within the phonon frequencies of the host, the localized mode of the defect losing energy by exciting phonons in the host. Such a mode is known as a *resonant band mode*, or *pseudo-localized mode*. This type of mode is illustrated in fig. 2.17(c) for a heavier defect, the frequency lowering with increased defect mass, but resonant modes can also occur for lighter defects if the force constants are sufficiently weak to allow the defect mode to lie within the phonon frequencies of the host. Resonant modes are predominantly found at the low frequency part of the host acoustic branches with heavy impurity masses. For a diatomic crystal if the defect atom is heavier than and replaces the lighter of the two types of atom then localized gap modes result, and if it is lighter than and replaces the heavier of the two types of atom then both localized modes and localized gap modes result.

Phonons and lattice vibrations

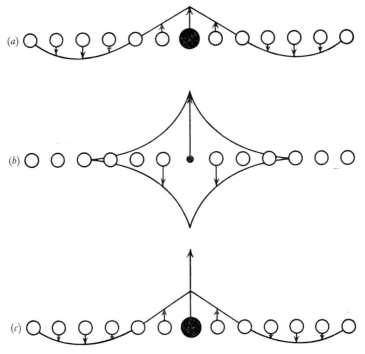

Fig. 2.17. Wave motion in an infinite linear monatomic lattice with a defect atom. Atoms of perfect crystal = ○; heavier defect atom = ●; lighter defect atom = •. (a) = Non-localized mode; (b) = localized mode; (c) = pseudo-localized mode.

2.8.5 The effect of polymeric linkages

When atoms or molecules in a crystal become linked together by strong chemical bonds to form polymeric chains many previously external vibrations become internal, and new models must be used in order to calculate the lattice dynamics.

Most studies have involved the use of a simple model using a planar zig–zag chain of atoms with cyclic boundary conditions. Using this model Kirkwood [189] used bending and stretching force constants to calculate the in-plane normal vibrations, Pitzer [190] included out of plane vibrations, which Krimm, Liang and Sutherland applied to polyethylene [191], and Tric [192] used the model of Kirkwood, but included two stretching force constants to account for single and double bonds. Schachtschneider and Snyder used computer methods [193] with a simple valence force field for polythene [194] and crystalline isotactic polypropylene [195], and their method has been extended into a general

Lattice dynamics of real crystals

method for the calculation of dispersion curves for any single polymer chain [196], which has been applied to inorganic systems [196a].

The simple model has been used to study chains of molecules hydrogen bonded together, and since the model considers interchain forces negligible compared with intramolecular forces, the number of molecules per unit cell is considered to be the number of chains per unit cell [197, 198]. Hydrogen bonded HF and DF chains have been studied [197], and the model extended to three dimensions [199].

One of the difficulties with using the simple model is that it takes no account of interchain forces. Tasumi and Shimanouchi [200] have considered interactions with neighbouring chains in a three-dimensional lattice and calculated intermolecular H—H force constants for crystalline polymethylene. In general very little is known about intermolecular force constants in polymers and the calculation of dispersion curves in the low frequency region becomes very difficult. In the low frequency region both internal torsional vibrations and external vibrations occur and these vibrations can be coupled to a remarkable extent, making the assignment of low frequency vibrations very difficult.

The vibrations of polymeric systems has been fully reviewed [201–6], and the simple model will now be used to examine the nature of these vibrations in zig–zag linear chains. In the hypothetical case of such one-dimensional chains motion about the axis for an isolated chain becomes an additional acoustic mode, which is not considered in the three-dimensional case. In real polymers 'end effects' have to be considered though little is known about the chain lengths when such effects become important. The following discussion will consider infinite polymers or polymers subject to cyclic boundaries. The vibrational ($K = 0$) modes of a zig–zag linear chain can be expressed in terms of the phase difference (ϕ) between successive atoms and the number of atoms in the unit cell (n) [191]

$$\phi = \frac{2\pi p}{n}, \qquad (2.22)$$

where p is an integer and is given by

$$p = 0, 1, \ldots, n-1.$$

Thus for $n = 2$, all $K = 0$ modes must have atoms with a phase difference of 0 or π (fig. 2.18) giving six modes. The dispersion relation for these six modes can be calculated by the Kirkwood and Pitzer method (see above), and is shown in fig. 2.19. Of the $3n$ modes, 3 are translations

Phonons and lattice vibrations

of the whole chain (for $K = 0$), and one a rotation of the whole chain (for $K = 0$). For $n = 2$ the masses of the two atoms in the unit cell can be equal since because of the zig–zag nature of the chain the smallest repeat unit is a unit of two atoms. For $n = 4$, $K = 0$ modes will have atoms

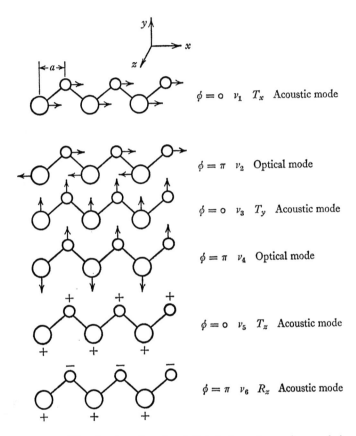

Fig. 2.18. Vibrational modes of an infinitely long planar zig–zag chain containing two atoms of different mass in the primitive cell.

with phase differences of 0, $\pi/2$, π, $3\pi/2$ giving twelve modes, the six new modes being illustrated in fig. 2.20. The dispersion relation can be obtained from fig. 2.19 by folding at a vertical line at $K = \pi/4a$ as explained in §2.3.4. In this case there will therefore be eight optical modes and four acoustic modes (three translations and one rotation).

Lattice dynamics of real crystals

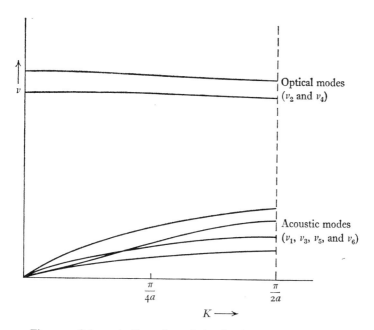

Fig. 2.19. Schematic dispersion relation for the modes in fig. 2.18.

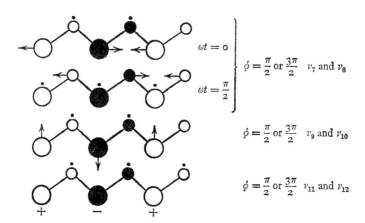

Fig. 2.20. Additional vibrational modes of an infinitely long planar zig–zag chain when it contains four atoms of different mass instead of two in the primitive cell. (• means the atom is at its equilibrium position.)

Phonons and lattice vibrations

2.8.6 *Force constants*

It is clear from the foregoing discussion that the calculation of force constants in real crystals involves large approximations. While such force constants are a necessary part of any simple models for calculating lattice dynamics, it is doubtful if the practice of converting experimental vibrational frequencies into force constants, whatever the approximations involved [207], has much value. Force constants calculated from experimental frequencies represent only one of many possible sets of numbers which when fed into some approximate potential energy formula will reproduce the observed frequencies. Only rarely is intermolecular interaction considered in such calculations.

2.9 EXPERIMENTAL METHODS FOR DETERMINING DISPERSION RELATIONS

2.9.1 *Specific heat measurements*

The contribution of phonon waves to the specific heats of solids has already been mentioned. The energy of each phonon wave can be expressed in terms of the kinetic and potential energy of the wave, and the energy of all the phonon waves added to give the total energy of the system, and hence the specific heat. In order to get a quantitative value for the temperature dependence of the specific heat it is necessary to know how the thermal energy distributes itself among the oscillators. Two models have been used for this purpose, one due to Einstein in 1907 and 1911, and one due to Debye in 1912. The details of these two models can be found elsewhere [31] and only the relevance of these models to real phonon waves will be discussed here.

Einstein considered a rather unrealistic model for a crystal, considering the atoms to vibrate independently of one another with a single frequency. The density of states for his model (the number of modes of vibration between ν and $\nu + \delta\nu$ given the symbol $g(\nu)$) therefore gives a straight line at this frequency, known as the Einstein frequency ν_E (fig. 2.21). While the Einstein model is rather unrealistic for all modes of a crystal it does approximate to the optical modes which generally show little dispersion (fig. 2.22).

Debye considered the modes of a crystal to be better considered as an elastic continuum, the dispersion relation for such elastic waves being that of fig. 2.2. The density of states for the Debye model is a con-

Experimental methods for dispersion relations

tinuous frequency spectrum with a cutoff frequency known as the Debye frequency ν_D (fig. 2.21). The Debye model therefore approximated to the acoustic modes of the crystal.

A combination of the two models might be expected to approximate the situation in real crystals. This is illustrated by comparing the density of states of a real crystal with a combination of the density of states for

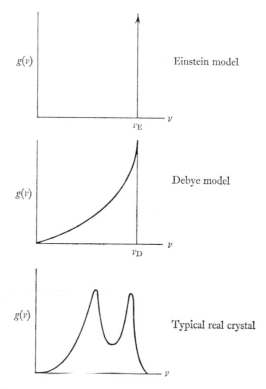

Fig. 2.21. Frequency distributions for the Einstein and Debye models, compared with that for a typical real crystal.

the two models (fig. 2.21). The optical modes corresponding to rotational and translational lattice modes would be expected to be given approximately by the Einstein or Debye frequency, and the Debye frequency has been used by some authors [208] for this purpose. Table 2.1 lists characteristic Debye frequencies for a variety of solids (from work by Clusius, Eucken, Gianque, Nernst, and Simon) listed in the textbook of Moelwyn-Hughes [209]. The agreement between the optical modes and

Phonons and lattice vibrations

the Debye frequency can be almost exact if comparison is made with the *centro-frequency*. The centro-frequency was defined by Plendl [210, 211]:

$$\nu_{\text{ctr}} = \int_{\nu_1}^{\nu_2} \nu f(\nu) \, d\nu \bigg/ \int_{\nu_1}^{\nu_2} f(\nu) \, d\nu, \qquad (2.23)$$

where $f(\nu)$ = reflectivity or absorption coefficient (e.g. for absorption it would be the extinction coefficient). The average is made over the entire range of the optical modes.

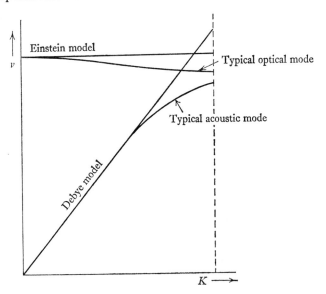

Fig. 2.22. Schematic dispersion relation for a linear infinite diatomic crystal compared with the dispersion relation for the Einstein and Debye models.

2.9.2 *Compressibility measurements*

In §2.3.1 it was seen that for elastic waves in an elastic band $\omega = v_0 K$. For acoustic modes the zone boundary frequency will be approximated by the intersection of the continuum expression of the elastic wave with the zone boundary (fig. 2.4 illustrates the degree of error in this approximation). For a linear lattice with n atoms in the unit cell

$$\nu = v_0/na = [na(\beta\rho)^{\frac{1}{2}}]^{-1}, \qquad (2.24)$$

where β = compressibility and ρ = density. Plendl [210, 212] has obtained good agreement between this frequency and the centro-frequency.

Experimental methods for dispersion relations

TABLE 2.1 *Debye frequencies for certain solids*

Solid	Debye frequency (cm^{-1})	Solid	Debye frequency (cm^{-1})
Li	269	F_2	69.6
Na	120	Cl_2	80.4
K	69.6	Br_2	60
Rb	40.3	I_2	53
Cs	29.4	$(CN)_2$	82.4
Cu	218	HF	124.8
Ag	149.5	HCl	89.4
Au	120.7	HBr	64.4
Be	697	HI	51
Mg	203	HCN	113
Ca	159	NaCl	196
Sr	118	KCl	161
Ba	80.4	KBr	124
Zn	164	H_2O	134.2
Cd	117.2	D_2O	128.6
Hg	67.7	H_2S	90.6
Al	272	N_2O	88.4
Tl	67	CO_2	97.6
C	1332	CS_2	67.4
Pb	61.3	COS	66.4
Fe	302.5	SO_2	83.6
Co	268.6	CaF_2	331
Ni	261.8	BaF_2	123.7
He	8.6	FeS_2	451
Ne	44	NH_3	146.7
Ar	58.7	$N(CH_3)_3$	99.7
Kr	44	PH_3	69
Xe	38.4	CH_4	54.4
H_2	63.7	CCl_4	59.4
N_2	47.4	$CsClO_4$	62
CO	55.6	N_2O_4	85.4
O_2	61	C_2H_6	91.3
NO	83	C_3H_6	69.6

Srivastava and Madan [213, 214] have discussed how the force constant g can be derived from an appropriate form of potential energy function (including van der Waals' forces, zero-point energy, and ionic polarization) and the compressibility and other lattice properties. The way in which the frequency can be obtained from g using dielectric constant data will be discussed in §4.3.

2.9.3 *Vibrational spectroscopic studies*

Vibrational spectroscopic studies can be carried out by infrared and Raman spectroscopic observations. Infrared spectroscopy is, in the first

Phonons and lattice vibrations

approximation, a simple direct process, the phonon or phonons excited having the same energy as the photon that excited it. In such an absorption process energy is removed from an incident monochromatic beam of incident photons by the destruction of photons and creation of phonons.

Raman spectroscopy is an inelastic scattering process. Monochromatic photons can either lose energy by being inelastically scattered by the creation of phonons in the crystal, or gain energy from phonons present

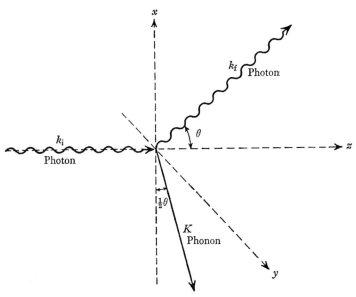

Fig. 2.23. Inelastic scattering of a photon of wave vector k_i, with the production of a phonon of wave vector K. The scattered photon has wave vector k_f. x, y, z refer to the principal axes of the crystal.

Conservation of energy: $h\nu = h\nu_t + h\nu'$.
Conservation of wave vector: $k_i = k_t + K$.

in the crystal. In this process the energy of the incident photon does not equal the energy of the created phonon or phonons, the remaining energy resulting in another photon of different frequency and different directions.

The detailed mechanisms of the infrared and Raman processes will be discussed in later chapters. For both processes energy and wave vector are conserved (§2.7). In the Raman process the phonon cannot take away much energy, since for any type of phonon the energy is much less than that of the photons usually used (u.v.). Fig. 2.23 illustrates the conservation of energy and wave vector in the inelastic scattering process.

Experimental methods for dispersion relations

2.9.4 Inelastic scattering of neutrons

Infrared and Raman spectroscopy is a limited tool for the experimental determination of dispersion relations because the first order spectra involve only zero phonon modes. The inelastic scattering of neutrons can provide a direct source of information about dispersion relations, since phonons of all wave vectors can be involved. The same information could in principle be deduced from the inelastic scattering of X-rays, but the great difference in the size of the X-ray and phonon quanta makes the method very difficult to apply. In the case of thermal neutrons the phonon and neutron energies are similar and the method has produced very useful results.

The major difference between inelastic scattering in the Raman effect and inelastic scattering in neutron studies lies in the scattering mechanism. In optical scattering spectra photons are scattered by molecular electron distributions and their movements, whereas neutrons are predominantly scattered by atomic nuclei. This difference in mechanism means that inelastic neutron scattering is not subject to the optical selection rules. In addition the compatibility of neutron and atomic masses allows the exchange of momentum as well as energy during a collision, leading to momentum transfers sufficient to span the entire Brillouin zone.

In the neutron inelastic scattering experiment a beam of monoenergetic neutrons of suitable energy is allowed to fall on the sample, and the beam scattered by collisions with the moving atomic nuclei in the sample. The neutron energy is determined and many neutrons will be found to have no energy change (elastic scattering), but some show a gain or loss of energy (inelastic scattering). Some neutrons scattered from the individual atoms will interfere, being scattered with phase angles definitely related to the types of site on which they are found; this is known as *coherent scattering*. Other neutrons scattered from the individual atoms will not interfere with each other; this is known as *incoherent scattering*. Neutron scattering spectra from predominantly incoherent scatterers (such as hydrogenous materials) and from predominantly coherent scatterers are the easiest to analyse. The study of coherent scattering spectra allows the phonons to be observed at a range of momentum transfers leading to the complete dispersion curve for both optical and acoustic phonons. It is important in the study of dispersion relations to minimize multiphonon scattering by a suitable choice of the experimental conditions. Conservation of energy and wave vector occurs and

Phonons and lattice vibrations

for one-phonon coherent scattering the conservation of wave vector is expressed by

$$k_f - k_i = G + K, \qquad (2.25)$$

where G is a reciprocal lattice vector, which arises because of inelastic scattering [215]. (2.25) is illustrated by fig. 2.24 which shows a section of the reciprocal lattice [216], the lattice points being separated by $1/a$, allowing wave vectors ($= 2\pi/\lambda$) to be drawn directly on the diagram. The experimental arrangement [218–21] allows the dispersion curves for each mode to be calculated with the aid of (2.25).

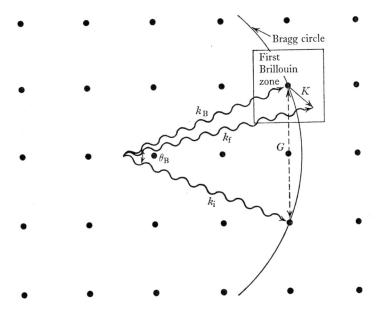

Fig. 2.24. Reciprocal lattice showing wave vectors in one-phonon coherent inelastic neutron scattering. k_i = wave vector of incident photon constructed so that one lattice point coincides with the end point of the vector. k_B = wave vector of elastically scattered photon at the Bragg angle θ_B to the incident photon, and lying on the Bragg circle. k_f = wave vector of inelastically scattered photon. K = wave vector of phonon. G = reciprocal lattice vector ($= \pi$ (actual reciprocal lattice vector) since $k = 2\pi/\lambda$).

Neutron incoherent inelastic scattering, while not providing information about dispersion relations, gives the density of states curves for the various phonon modes. Such scattering always occurs when mono-energetic neutrons fall on a sample containing hydrogen, and has proved extremely useful in investigating phonon modes involving hydrogen atoms, particularly since the neutron inelastic scattering cross-section of

Experimental methods for dispersion relations

hydrogen is an order of magnitude greater than for most other nuclei. Therefore experiments involving hydrogen containing samples yield an observed spectrum that primarily reflects the motion of the hydrogen atoms, and in addition is not subject to the selection rules that restrict the observed modes in optical spectra.

Cold-neutron scattering experiments involve the neutrons gaining energy from thermal phonons present in the crystal and correspond to anti-Stokes scattering in Raman spectroscopy. Higher energy neutron scattering experiments, with the neutrons losing energy to form phonons, correspond to Stokes scattering in Raman spectroscopy. Neutron incoherent inelastic scattering has been used to study organic [219, 220, 222–5], inorganic [218, 219, 226, 227] and metallic [227c] systems.

The full details of the theory of inelastic neutron scattering, together with a full discussion and examples can be found elsewhere [218–21].

3 The application of group theory to a crystal lattice

3.1 INTRODUCTION

In the previous chapter it was seen how lattice vibrations depend upon the periodicity of a crystal lattice. Group theory is now applied to the crystal lattice in order that this periodicity can be clearly defined in particular lattices. Once the periodicity of a particular crystal lattice is so defined the normal modes active in infrared and Raman studies can then be calculated.

Group theory has been the subject of numerous books and review articles, and most books concerned with the theory of the solid state have chapters on group theory. Excellent introductions to the subject are given in the books by Cotton [228] and Schonland [229]. Reviews of the use of group theory in crystal physics have been given by Johnston [230] and Jones [231], and the detailed treatment of group theory in quantum mechanical systems has been given by Heine [232] and Tinkham [233]. The application of group theory to molecular vibrations has been thoroughly discussed by Wilson, Decius and Cross [234]. Mathieu [7] and Bhagavantam and Venkatarayudu [235] have considered the application of group theory to vibrations in crystals, and Hornig [1] and Winston and Halford [236, 237] developed the site and unit cell group approach discussed in this chapter. Zbinden [201] gives a useful discussion of group theory and its application to polymer vibrations.

In this chapter the group theory necessary to define the periodicity of a particular lattice in terms that will allow spectral activities to be discussed will be outlined, and the reader is referred to the above references for a complete discussion.

3.2 GROUPS

3.2.1 *The Group*

A *group* is a collection of abstract entities called *elements*. Discussion of the group requires these elements to be neither specified nor assigned any physical significance. In order for any set of elements to form a mathematical group they must satisfy the following conditions:

Groups

(i) *AB* is an element of the group whenever *A* and *B* are;
(ii) *ABC* = *A*(*BC*) = (*AB*)*C* for all *A*, *B*, *C*, in group;
(iii) there is an element *E* of the group such that

$$EX = XE = X$$

for all *X* in group;
(iv) for every element *R* of the group, there is an R^{-1} such that

$$RR^{-1} = R^{-1}R = E.$$

If *AB* = *BA* for all *A*, *B* in group, then the group is an *Abelian Group*. *E* is called the *identity element*, and R^{-1} is the reciprocal of *R*. Groups can be finite or infinite, and the number of elements in the group, *h*, is called the *order of the group*.

Cyclic groups are Abelian groups with elements X, X^2, X^3, X^4, ..., X^n such that $X^n = E$.

Isomorphic groups are groups in which every element of one group corresponds to an element of the other group and vice versa.

Subgroups occur when part of the elements of a group form a group on their own which is called a subgroup of the original group. The order of any subgroup, *g*, must be a divisor of the order of the group, i.e. *h*/*g* = *l*, where *l* is an integer called the 'index' of the subgroup.

3.2.2 *The class*

Elements of a group can be separated into smaller sets of elements either as a subgroup (above), or as a class. A class consists of a set of elements that are conjugate to one another. An element conjugate to *A* is defined as $X^{-1}AX$, where *X* is any element of the group, therefore if one element of the class is given, all others can be obtained by conjugation with every element of the group. The definition of class leads to further definitions. A *self-conjugate subgroup* (*I*) (sometimes called invariant subgroup) is a subgroup which consists of full classes. A *coset* is a set of elements obtained by the multiplication of each element of a subgroup with one element of the group. In the case of a self-conjugate subgroup the cosets are of particular importance since the complexes obtained by left or right multiplication are identical:

$$A(I) = (I)A$$

where *A* = element of the group, and (*I*) the collection of elements that make up the self-conjugate subgroup.

Group theory applied to crystal lattice

3.2.3 *The factor group*

The factor group is defined as a group (F) whose elements are cosets of a self-conjugate subgroup, i.e. each element will be represented by $A(I)$ or $(I)A$. The order of the factor group is therefore equal to the number of non-equivalent cosets of the self-conjugate subgroup which will be the 'index' l of (I).

3.3 THE SPACE GROUP

The theory of space groups [238–40] shows that a lattice structure must have the symmetry of one of the 230 space groups. The symmetry operations involved include the translations which generate the lattice plus the symmetry operations familiar in point groups. The *unit cell* is taken to be the smallest unit in which no atoms are equivalent under simple translations, though they may be equivalent under any other symmetry operations. The complete set of symmetry operations for the space group of the crystal can thus be obtained from those of the unit cell by these translations. The choice of the unit cell is not unique, which makes no difference in the discussion, though the smallest volume unit cell is needed for the analysis of vibrational modes.

The symmetry operations of the space group will now be considered. First the symmetry operations familiar in point groups will be discussed. All the point group symmetry operations can be regarded either as simple rotations, or as rotation–reflections. It can be shown that there are only 11 possible pure rotational point groups [241]:

$$C_1, C_2, C_3, C_4, C_6, D_2, D_3, D_4, D_6, T, O, \tag{3.1}$$

the notation being given in the Schönflies system. All space groups explicitly possess a centre of inversion as a symmetry operation, and so this must be included as a crystallographically possible operation. To do this the 11 pure rotational point groups are multiplied in turn with the group $\bar{1}$ which consists of the elements E and i which will give 11 mixed groups:

$$C_i, C_{2h}, C_{3i}, C_{4h}, C_{6h}, D_{2h}, D_{3d}, D_{4h}, D_{6h}, T_h, O_h. \tag{3.2}$$

The order of these new groups will clearly be twice that of the pure rotation group, the elements of the new groups being written out by combining E and i with each of the elements of the rotation group in turn. All the point group symmetry operations have now been considered

The space group

and so the possible point groups will be those given in (3.2) and those in all the distinct subgroups of (3.2). The 11 pure rotation groups in (3.1) are one set of such subgroups, and a further set of ten more subgroups, which are mixed groups that do not possess i explicitly, can be obtained by examination:

$$C_s, C_{2v}, S_4, C_{4v}, D_{2d}, C_{3v}, D_{3h}, C_{6v}, T_d. \quad (3.3)$$

There are thus 32 possible point groups and these are the groups listed in (3.1), (3.2) and (3.3).

It can be shown that all space groups can be classified by seven crystal systems which are characterized by 7 point groups:

$$C_i, C_{2h}, D_{2h}, D_{4h}, D_{3d}, D_{6h}, O_h. \quad (3.4)$$

The crystal systems, together with the 32 point groups and their elements of symmetry are listed in table 3.1.

Secondly the translation symmetry operations must be considered. These operations, which generate the lattice, consist of translations and operations involving translations such as screw axes and glide planes. So far the space group has been considered to be of infinite order, but real crystals are of finite size so that a finite space group is considered. This is equivalent to adopting Born–von Kármán cyclic boundary conditions. The translation operations will now be considered as a subgroup of the finite space group.

3.4 SUBGROUPS OF THE FINITE SPACE GROUP

3.4.1 *The translation group*

The translational operations of the space group form a subgroup of the finite space group known as the *translation group*. An element of the translation group, T, will be

$$t_{n_1 n_2 n_3} = n_1 t_1 + n_2 t_2 + n_3 t_3, \quad (3.5)$$

where t_1, t_2, t_3 represent primitive translation vectors that move one atom or molecule into the next identical one, and n_1, n_2 and n_3 are integers that may be positive, negative, or zero. The order of the finite translation group is N_1, N_2, N_3, where N_1, N_2, N_3 represent upper limits to the values of n_1, n_2 and n_3. The group is an Abelian group since the product of any two elements is a summation of two vectors, and such a summation is commutative. The group is also a self-conjugate subgroup since it contains full classes.

Group theory applied to crystal lattice

TABLE 3.1 *Point groups*

Classification on basis of optical refraction		Systems	Class International	Class Schönflies	Symmetry elements
Anisotropic	Biaxial	Triclinic	1	C_1	E
			$\bar{1}$	C_i	$E\,i$
		Monoclinic	m	C_s	$E\,\sigma_h$
			2	C_2	$E\,C_2$
			$2/m$	C_{2h}	$E\,C_2\,i\,\sigma_h$
		Orthorhombic	$2mm$	C_{2v}	$E\,C_2\,\sigma'_v\,\sigma''_v$
			222	D_2	$E\,C_2\,C'_2\,C''_2$
			mmm	D_{2h}	$E\,C_2\,C'_2\,C''_2\,i\,\sigma_h\,\sigma'_v\,\sigma''_v$
		Tetragonal	4	C_4	$E\,2C_4\,C_2$
			$\bar{4}$	S_4	$E\,2S_4\,C_2$
			$4/m$	C_{4h}	$E\,2C_4\,C_2\,i\,2S_4\,\sigma_h$
			$4mm$	C_{4v}	$E\,2C_4\,C_2\,2\sigma'_v\,2\sigma_d$
			$\bar{4}2m$	D_{2d}	$E\,C_2\,C'_2\,C''_2\,2\sigma_d\,2S_4$
			422	D_4	$E\,2C_4\,C_2\,2C'_2\,2C''_2$
			$4/mmm$	D_{4h}	$E\,2C_4\,C_2\,2C'_2\,2C''_2\,i\,2S_4\,\sigma_h\,2\sigma'_v\,2\sigma_h$
	Uniaxial	Rhombohedral (Trigonal)	3	C_3	$E\,2C_3$
			$\bar{3}$	S_6	$E\,2C_3\,i\,2S_6$
			$3m$	C_{3v}	$E\,2C_3\,3\sigma_v$
			32	D_3	$E\,2C_3\,3C_2$
			$\bar{3}m$	D_{3d}	$E\,2C_3\,3C_2\,i\,2S_6\,3\sigma_d$
		Hexagonal	$\bar{6}$	C_{3h}	$E\,2C_3\,\sigma_h\,2S_3$
			6	C_6	$E\,2C_6\,2C_3\,C_2$
			$6/m$	C_{6h}	$E\,2C_6\,2C_3\,C_2\,i\,2S_3\,2S_6\,\sigma_h$
			$\bar{6}m2$	D_{3h}	$E\,2C_3\,3C_2\,\sigma_h\,2S_3\,3\sigma_v$
			$6mm$	C_{6v}	$E\,2C_6\,2C_3\,C_2\,3\sigma_v\,3\sigma_d$
			622	D_6	$E\,2C_6\,2C_3\,C_2\,3C'_2\,3C''_2$
			$6/mmm$	D_{6h}	$E\,2C_6\,2C_3\,C_2\,3C'_2\,3C''_2\,i\,2S_3\,2S_6\,\sigma_h\,3\sigma_d\,3\sigma_v$
Isotropic		Cubic	23	T	$E\,4C_3\,4C_3^2\,3C_2$
			$m3$	T_h	$E\,4C_3\,4C_3^2\,3C_2\,i\,8S_6\,3\sigma_h$
			$\bar{4}3m$	T_d	$E\,8C_3\,3C_2\,6\sigma_d\,6S_4$
			432	O	$E\,8C_3\,3C_2\,6C'_2\,6C_4$
			$m3m$	O_h	$E\,8C_3\,3C_2\,6C_2\,6C_4\,i\,8S_6\,3\sigma_h\,6\sigma_d\,6S_4$

3.4.2 *The factor group*

Let there be H point group symmetry operations of the finite space group. The finite space group will thus have an order $N_1N_2N_3H$. The translation group therefore has an index H. The *factor group* of the space group will therefore be the group whose elements are cosets of the translation group (a self-conjugate subgroup), and will

Space group and subgroup elements

be of order H. The H cosets can be explicitly written down in the following form:

$$
\left.\begin{aligned}
F_1 &= Et_{001}, Et_{002}, \ldots, Et_{N_1 N_2 N_3}, \\
F_2 &= P_2 t_{001}, P_2 t_{002}, \ldots, P_2 t_{N_1 N_2 N_3}, \\
&\vdots \\
F_H &= P_H t_{001}, P_H t_{002}, \ldots, P_H t_{N_1 N_2 N_3},
\end{aligned}\right\} \quad (3.6)
$$

where P represents a point group symmetry element of the space group ($P_1 = E$).

3.4.3 *The unit cell group*

This group describes the symmetry of the unit cell. It is defined as a finite subgroup of the finite space group if translations which carry a point in a unit cell into the equivalent point in another unit cell are defined as identity. It will clearly have the order H.

There has been a certain amount of discussion in the recent literature [217, 242, 243] as to the exact definition of the factor and unit cell groups, and the situation has recently been clarified by Bertie and Bell [243]. The definitions given here are those suggested by Bertie and Bell, and the distinction between the unit cell and factor groups should be clear when the elements of these groups are considered in the next section.

3.4.4 *The site group*

A site is defined as a point which is left invariant by some operations of the space group. These operations form a subgroup of the finite space group called a *site group*. It follows that every point is a site, having at least the trivial site group C_1.

3.4.5 *The interchange group*

All the elements of the unit cell group *not* included in the site group have been called the *interchange elements* {244, 242}. The *interchange group* has been defined as the group containing these elements and the identity element.

Group theory applied to crystal lattice

3.5 ELEMENTS OF THE FINITE SPACE GROUP AND ITS SUBGROUPS

Consider a space group S. The space group will have the elements:

$$\left.\begin{aligned}S_1 &= Et_{001},\\ S_2 &= Et_{002},\\ &\vdots\\ S_{N_1N_2N_3} &= Et_{N_1N_2N_3},\\ S_{N_1N_2N_3}+1 &= P_2 t_{001},\\ S_{N_1N_2N_3}+2 &= P_2 t_{002},\\ &\vdots\\ S_{N_1N_2N_3H} &= P_H t_{N_1N_2N_3}.\end{aligned}\right\} \quad (3.7)$$

The factor group has the elements given in (3.6). The unit cell group (U) will consist of the elements:

$$\left.\begin{aligned}U_1 &= Et_{000},\\ U_2 &= P_2 t_{000},\\ U_3 &= P_3 t_{000},\\ &\vdots\\ U_H &= P_H t_{000}.\end{aligned}\right\} \quad (3.8)$$

The site group will contain all elements which leave the site invariant, and involves only point symmetry operations. For some site group Si^x with point symmetry elements $P_{x_1}, P_{x_2}, P_{x_3}$ the elements will be

$$\left.\begin{aligned}Si^x_1 &= Et_{000},\\ Si^x_2 &= P_{x_1} t_{000},\\ Si^x_3 &= P_{x_2} t_{000},\\ Si^x_4 &= P_{x_3} t_{000}.\end{aligned}\right\} \quad (3.9)$$

The difference between the space group and its subgroups and between the subgroups should now be clear, since they all contain different numbers of elements, or different types of elements (unit cell group and factor group). Table 3.2 lists the order of these groups.

3.5.1 *Symmorphic and non-symmorphic space groups*

In the foregoing discussion only elements of symmorphic space groups have been considered. Non-symmorphic space groups contain screw

Elements of finite space group

TABLE 3.2 *Orders of the space group and its subgroups*

Group or subgroup	Order
Infinite space	Infinite
Finite space	$N_1 N_2 N_3 H$
Finite translation	$N_1 N_2 N_3$
Factor	H
Unit cell	H
Site	H/n
Interchange	$H - H/n$

n = number of equivalent sites in a set.

axes and glide planes. The presence of screw axes and glide planes means that they must be represented by operations involving a specific combination of primitive translations. For example a screw axis with a translation in the b direction would be represented by $C_2 t_{0\frac{1}{2}0}$ and a glide plane with a translation in the c direction would be represented by $\sigma_h t_{00\frac{1}{2}}$. Consider a non-symmorphic space group S', which contains the screw axis $C_2 t_{0\frac{1}{2}0}$ and the glide plane $\sigma_h t_{00\frac{1}{2}}$ replacing point group symmetry elements P_2 and P_3. The elements of S' will be those of S (3.7) with the following modified elements:

$$\left. \begin{aligned} S_{N_1 N_2 N_3 + 1} &= C_2 t_{0\frac{1}{2}0} t_{001}, \\ &\vdots \\ S_{2 N_1 N_2 N_3} &= C_2 t_{0\frac{1}{2}0} t_{N_1 N_2 N_3}, \\ S_{2 N_1 N_2 N_3} &= \sigma_h t_{00\frac{1}{2}} t_{001}, \\ &\vdots \\ S_{3 N_1 N_2 N_3} &= \sigma_h t_{00\frac{1}{2}} t_{N_1 N_2 N_3}. \end{aligned} \right\} \qquad (3.10)$$

The factor group, F', of S' will contain the modified elements of S' (3.10) in its elements:

$$\left. \begin{aligned} F'_2 &= C_2 t_{0\frac{1}{2}0} t_{001}, C_2 t_{0\frac{1}{2}0} t_{002}, \ldots, C_2 t_{0\frac{1}{2}0} t_{N_1 N_2 N_3}, \\ F'_3 &= \sigma_h t_{00\frac{1}{2}} t_{001}, \sigma_h t_{00\frac{1}{2}} t_{002}, \ldots, \sigma_h t_{00\frac{1}{2}} t_{N_1 N_2 N_3}, \end{aligned} \right\} \qquad (3.11)$$

and the unit cell group, U', will contain the modified elements of S' in its elements:

$$\left. \begin{aligned} U'_2 &= C_2 t_{0\frac{1}{2}0} t_{000}, \\ U'_3 &= \sigma_h t_{00\frac{1}{2}} t_{000}. \end{aligned} \right\} \qquad (3.12)$$

The site groups will be unaffected since they involve only point group symmetry operations.

Group theory applied to crystal lattice

In the case of symmorphic space groups the elements of the factor and unit cell groups can be seen to be isomorphic to one of the thirty-two crystallographic point group elements, P. For non-symmorphic space groups some of the elements P must be replaced by non-point group screw axes and glide planes (e.g. C_2 is replaced by $C_2 t_{0\frac{1}{2}0}$). International tables [245] list the 230 space groups, and give a diagram of the unit cell. These diagrams show all the operations of the factor group. In the Schönflies system the symbols (which can be taken to represent the factor group as well as the space group) are identical to the symbols for point groups with the addition of a superscript. The superscript is necessary since for a given point group, a number of possibilities arise by adding elements of screw axes and glide planes in a number of different ways. Thus for the point group D_2 (International class symbol 222) nine possibilities arise. In the first the site and unit cell group symmetry is D_2, there being no screw axes or glide planes. This space group is denoted D_2^1 (International class symbol P 222). In the second a screw rotation is introduced, and the site symmetry is lowered through the loss of the point group element that becomes the screw axis, and the space group is denoted D_2^2 (International symbol P 222_1). Variations in the introduction of screw axes and/or glide planes lead to D_2^3 to D_2^9.

Halford [237] has listed the possible site groups for the various space group symmetries, and Adams and Newton [246, 247] have recently corrected this list.

3.6 SPACE GROUPS DURING LATTICE VIBRATIONS

The symmetry elements of the crystal lattice have now been investigated in terms of the group and subgroups they can form. The dependence of particular lattice vibrations on the periodicity of the crystal lattice can now be discussed. In order to do this the displacements of the atoms and molecules from their equilibrium positions must be expressed in terms of their effect on the symmetry elements of the crystal lattice. This can be achieved by associating with each element X a square matrix $M(X)$. The set of matrices $M(E), M(X^1), M(X^2), \ldots$ that forms a group isomorphic to the group $E, X^1, X^2, \ldots$ is called the *representation* of the group. The elements $M(X)$ of the representation must satisfy the characteristic group relations that hold for the elements X of the group. Representations can now be written which show how the wave vector representing the

Space groups during lattice vibrations

motion of the atoms and molecules from their equilibrium positions is affected by the symmetry of the crystal lattice.

There is no restriction on the dimension of the matrices $M(X)$, but those matrices which form the *reducible representation* of the group can be reduced to form an *irreducible representation* of the group. This reduction is performed by applying a transformation matrix T to $M(X)$:

$$TM(X)T^{-1} = M'(X), \tag{3.13}$$

where T and $M(X)$ have the same dimension. If T can be found so that $M'(X)$ has the form

$$M'(X) = \begin{bmatrix} M'(X)_1 & & 0 \\ & \ddots & \\ 0 & & M'(X)_n \end{bmatrix}_n \tag{3.14}$$

then $M'(X)$ can be further reduced. If $M'(X)_1$ to $M'(X)_n$ cannot be further reduced by some T then they will be the elements of $M(X)$ in the irreducible representation of the group. The *character* of a square matrix is the sum of its diagonal elements and is given the symbol χ. These characters rather than the complete matrices can be used for many important deductions.

Consider an arbitrary wave vector K describing a particular lattice vibration. K can be expressed in terms of the reciprocal lattice vectors t_1^*, t_2^*, t_3^* which describe the translational symmetry of the lattice ($|t_1^*| = 1/|t_1|$ etc.), since $K = 2\pi/\lambda$. Thus for a general three-dimensional lattice

$$K = \pi \left[\frac{x_1}{N_1} t_1^* + \frac{x_2}{N_2} t_2^* + \frac{x_3}{N_3} t_3^* \right], \tag{3.15}$$

where x_1, x_2 and x_3 are integers running from 0 to N_1, N_2 and N_3 respectively (this follows from the definitions of the Brillouin zone in the previous chapter). The Brillouin zone will therefore be a unit cell of the reciprocal lattice (fig. 2.24). A discussion of the choice of this unit cell can be found elsewhere [248]. K will be mapped into an assembly of different vectors (they have the same magnitude, though they point in different directions) by the point group symmetry elements of the space group. For one such element P the process

$$K \xrightarrow{\text{Operation } P} K_P \tag{3.16}$$

must be considered. All the vectors K_P that are identical with K are placed in the *group of K* [236, 249], represented by $[K]$, which therefore

Group theory applied to crystal lattice

contains all elements P which leave a given K invariant. All the vectors K_P that are different from K (i.e. point in different directions) are placed in another group, called the *star of K*, represented by $\{K\}$, which contains all the assembly of vectors into which K is mapped (i.e. K_P) by P, not contained in $[K]$. The star of K therefore defines the irreducible representation of the finite translation group in three dimensions. The general irreducible representation of the space group S'' of the crystal lattice whose symmetry is modified by the presence of the lattice vibrations is therefore defined by

$$S'' = \{K\}+[K]. \tag{3.17}$$

Thus the subgroups of the space group S, the finite translation and unit cell groups defined by

$$S = T+U \tag{3.18}$$

are replaced by the star and the group of K respectively when S'' is considered. The orders of these two subgroups vary depending upon the value of K, thus if the group of K has an order J, the star of K will have an order H/J (since there are H elements P).

The elements of S'' described in terms of the atomic and molecular positions will be represented by square matrices $M(X)$. These representations Γ will generally be reducible to some irreducible representation Γ_k. It can be shown (see the references in §3.1) that the number of times that Γ_k is contained in Γ is given by

$$n_k = \frac{1}{Z}\sum_j h_j \chi_k^*(P)\chi_j'(P), \tag{3.19}$$

where Z is the order of the group or subgroup, h_j is the number of group operations in the jth class, and $\chi_j'(P)$ and $\chi_k(P)$ are the characters of the representations Γ and Γ_k. In order to discover the number of different modes for a single K value the number of modes belonging to different irreducible representations of S'' are required. Tables that list characters of representations are known as *character tables* and the methods for calculating character tables of the elements of S'' for the first Brillouin zone can be found elsewhere [249].

A special case arises for which the situation becomes appreciably simpler. This is the case when $K = 0$. In this case all elements P leave K invariant, and therefore the star of K contains one wave vector K_1 which is the effect of the identity operaton on K and so is identical to K (sometimes it is called the generating vector of the star [249]). The group of K will now be of order H and will be identical to the unit cell group whose

Space groups during lattice vibrations

symmetry is modified by the presence of the lattice vibrations, for the *smallest volume unit cell*. The star of K will therefore represent a totally symmetric representation of the translation group, T^s. Thus for $K = 0$

$$S'' = T^s + U. \qquad (3.20)$$

The product of the individual elements of two subgroups of a group is not generally commutative, but it is commutative when one subgroup is totally symmetric. Thus the product of the elements of U and T^s will give the elements of S''. The number modes belonging to different irreducible representations of S'' will thus be obtainable from the character tables of the unit cell group since T^s has one point group element E. The character tables for the unit cell group will be the point group character tables (appendix) bearing in mind that for non-symmorphic space groups non-point group operations are involved.

3.7 UNIT CELL GROUP ANALYSIS

In the previous section it was seen how for $K = 0$ the types of modes of the crystal could be calculated from an analysis of the unit cell of smallest volume. Many crystallographic unit cells contain several unit cells of the smallest volume so that before an analysis can be made the smallest volume unit cell must be determined.

Consider an atom or molecule of the crystal and let it be displaced a distance x from its equilibrium position during a lattice vibration. If a coordinate is defined so that it lies along the direction of x, this coordinate will be called a *normal coordinate* Q_k, and the motion said to be a *normal vibration*, or *normal mode*. In general, normal coordinates will consist of a complicated combination of atomic or molecular displacements, and will be linear combinations of the original coordinates. To find the types of normal modes for $K = 0$ the characters of the reducible representations of the symmetry elements of the unit cell group whose symmetry is modified by that mode are written down. The characters of the reducible representations of the unit cell group are given by the character table for the group and so the number of types of normal mode can be calculated using (3.19). The procedure for different types of normal mode will now be discussed.

3.7.1 *All modes*

In order to determine the total number of unit cell modes belonging to each irreducible representation of the unit cell group, each atom in the

Group theory applied to crystal lattice

smallest volume unit cell is considered. The total number of atoms invariant under each symmetry operation, $\mathscr{U}_x$, represented by the elements of the unit cell group U_x, is counted. The atoms are considered invariant if they are either left in their initial position or carried to an identical position in another unit cell, and each invariant atom contributes 1 to a quantity ω_{U_x} which represents the sum for each of the unit cell elements $U_1, U_2, \ldots, U_H$. The character of the reducible representation for all the modes, $\chi'_j(n_i)$ will be given by

$$\chi'_j(n_i) = \sum_x \omega_{U_x}(\pm 1 + 2\cos\phi_{U_x}), \qquad (3.21)$$

where ϕ_{U_x} is the angle of rotation corresponding to the symmetry operation U_x, the plus and minus signs standing respectively for proper and improper rotations. $(\pm 1 + 2\cos\phi_{U_x})$ represents the contribution to the character per unshifted atom, and values for different symmetry operations U_x are listed in table 3.3. The total ω_{U_x} for all the unit cell elements is given by ω_U where

$$\omega_U = \sum_x \omega_{U_x}. \qquad (3.22)$$

TABLE 3.3 *Contributions to character per unshifted atom*

Proper rotations		Improper rotations	
Symmetry operation ($\mathscr{U}_x$)	χ	Symmetry operation ($\mathscr{U}_x$)	χ
C_n^k	$1 + 2\cos\phi_{U_x}$	S_n^k	$-1 + 2\cos\phi_{U_x}$
$E = C_1^k$	3	$\sigma = S_1^1$	1
C_2^1	-1	$i = S_2^1$	-3
$C_3^1 \quad C_3^2$	0	$S_3^1 \quad S_3^5$	-2
$C_4^1 \quad C_4^3$	1	$S_4^1 \quad S_4^3$	-1
$C_6^1 \quad C_6^5$	2	$S_6^1 \quad S_6^5$	0

3.7.2 External modes

(i) Acoustic modes

These three modes must be subtracted from the total number of modes (see previous chapter). The types of these modes can be found from the character of the reducible representation for these modes, $\chi'_j(T)$, which will be given by

$$\chi'_j(T) = \sum_x (\pm 1 + 2\cos\phi_{U_x}) \qquad (3.23)$$

since $\omega_U = 1$ because the unit cell moves as one unit in these modes.

(ii) Translational modes

Consider the m molecular groups in the unit cell in the same way as the atoms were considered in (3.21). The character of the reducible representation for these modes will be given by

$$\chi'_j(T') = [\sum_x \omega_{U_x}(m) - 1] \sum_x (\pm 1 + 2\cos\phi_{U_x}) \quad (3.24)$$

where the acoustic modes have been subtracted. $\omega_{U_x}(m)$ represents the number of molecular groups invariant for the unit cell element U_x for translational modes.

(iii) Rotational modes

(a) Non-linear molecules. Let v of the m molecular groups have no degrees of rotational freedom. The remaining $(m-v)$ groups are considered in the same way as before. The character of the reducible representation for these rotational modes will be given by

$$\chi'_j(R') = \sum_x \omega_{U_x}(m-v)(\pm 1 + 2\cos\phi_{U_x}), \quad (3.25)$$

where $\omega_{U_x}(m-v)$ represents the number of molecular groups capable of rotation invariant for the unit cell element U_x for rotational modes.

(b) Linear molecules. In linear molecules the number of degrees of freedom is reduced from three to two. In this case table 3.3 must be modified since contributions to the character involve 2×2 matrices rather than 3×3 matrices. For such a case it can be shown that the contribution to the character per unshifted group for rotation about the molecular axis is $\mp 2\cos\phi_{U_x}$, hence 1 must be subtracted from each of the contributions in table 3.3. Rotations perpendicular to the molecular axis will contribute zero to the character (the cross terms of the 2×2 matrix add to zero), and such rotations will include C_2 axes and σ_v planes. The character of the reducible representation for these rotational modes will be given by

$$\chi'_j(R') = \sum_x \omega_{U_x}(m-v)(\pm 2\cos\phi_{U_x})$$

for $U_x = C(\phi)$ or $S(\phi)$, and

$$\chi'_j(R') = 0 \quad (3.26)$$

for $U_x = C_2$ or σ_v.

Group theory applied to crystal lattice

3.7.3 Internal modes

In order to determine the number of internal modes belonging to each irreducible representation of the unit cell group (n'_i) the sum of the acoustic, translational, and rotational modes is subtracted from the total number (n_i).

3.7.4 Illustration

The analysis described above will now be illustrated with an example. A number of substances, such as calcite ($CaCO_3$) [250] and the high temperature phase of KNO_3 (phase I) [251] have the D_{3d}^6 space group.

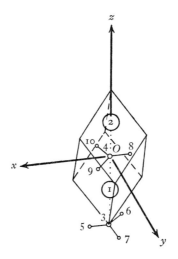

Fig. 3.1. Unit cell of calcite ($CaCO_3$) or KNO_3 (phase I).

For the two examples given the unit cell of smallest volume is rhombohedral and contains two molecules. This is always the first step and often the nature of the unit cell is known from X-ray crystallographic studies, though in some cases the unit cell from such studies is not the smallest volume unit cell. For the examples chosen $n = 10$, $m = 4$ and $v = 2$, and the unit cell is illustrated in fig. 3.1.

Table 3.4 lists the effect that the symmetry operations represented by the unit cell group elements U_x have on the atoms in the unit cell. From tables 3.3 and 3.4, (3.21), (3.23), (3.24), (3.25), and (3.26) can be evaluated, and so the number of the various modes belonging to each

Unit cell group analysis

TABLE 3.4 *The effect of symmetry operation $\mathcal{U}_x$ on unit cell atoms*

$\mathcal{U}_x$		Atom sites invariant under $\mathcal{U}_x$	Atom sites variant under $\mathcal{U}_x$	
			Initial site	Final site after $\mathcal{U}_x$
E Identity		1, 2, 3, 4, 5, 6, 7, 8, 9, 10		
$2C_3$ Rotation $\pm 2\pi/3$ about OZ axis	$C_3(1)$	1, 2, 3, 4	8, 9, 10	9, 10, 8
			5, 6, 7	6, 7, 5
	$C_3(2)$	1, 2, 3, 4	8, 10, 9	10, 9, 8
			5, 7, 6	7, 6, 5
$3C_2$ Rotation π about OX or an axis deduced by C_3	$C_2(1)$	3, 4, 5, 8	1, 2	2, 1
			6, 7	7, 6
			9, 10	10, 9
	$C_2(2)$	3, 4, 6, 9	1, 2	2, 1
			5, 7	7, 5
			8, 10	10, 8
	$C_2(3)$	3, 4, 7, 10	1, 2	2, 1
			5, 6	6, 5
			8, 9	9, 8
i Inversion		1, 2	3, 4	4, 3
			5, 8	8, 5
			6, 9	9, 6
			7, 10	10, 7
$2S_6$ Rotation $\pm 2\pi/6$ about OZ axis with reflection through a plane perpendicular to OZ	$S_6(1)$	1, 2	3, 4	4, 3
			5, 9	9, 5
			7, 8	8, 7
			6, 10	10, 6
	$S_6(2)$	1, 2	3, 4	4, 3
			5, 10	10, 5
			6, 8	8, 6
			7, 9	9, 7
$3\sigma_d$ (glide) Reflection through ZOY or plane deduced by C_3, with translation $= \frac{1}{2}(a+b+c)$	$\sigma_d(1)$		1, 2	2, 1
			3, 4	4, 3
			5, 8	8, 5
			6, 10	10, 6
			7, 9	9, 7
	$\sigma_d(2)$		1, 2	2, 1
			3, 4	4, 3
			5, 10	10, 5
			6, 9	9, 6
			7, 8	8, 7
	$\sigma_d(3)$		1, 2	2, 1
			3, 4	4, 3
			5, 9	9, 5
			6, 8	8, 6
			7, 10	10, 7

Group theory applied to crystal lattice

irreducible representation of the unit cell group, n_i, T, T', R', and n'_i can be found by using the unit cell group character table (table 3.5). As a result of the character 2 for the E element of the irreducible representations E_1 and E_2 the vibrations belonging to these irreducible representations are degenerate. Of the 30 modes, there are nine single modes, nine doubly degenerate modes, and one single and one doubly degenerate acoustic modes. The Raman and infrared activities will be discussed later.

TABLE 3.5 *Character table for the unit cell group of the D_{3d}^6 space group and the numbers of various modes for $CaCO_3$ or KNO_3 (phase I)*

D_{3d}	Γ	E	$2C_3$	$3C_2$	i	$2S_6$	$3\sigma_d$(glide)	n_i	T	T'	R'	n'_i	R	I
A_{1g}	1	1	1	1	1	1	1	1	0	0	0	1	a	f
A_{2g}	2	1	1	−1	1	1	−1	3	0	1	1	1	f	f
E_g	3	2	−1	0	2	−1	0	4	0	1	1	2	a	f
A_{1u}	4	1	1	1	−1	−1	−1	2	0	1	0	1	f	f
A_{2u}	5	1	1	−1	−1	−1	1	4	1	1	1	1	f	a
E_u	6	2	−1	0	−2	1	0	6	1	2	1	2	f	a
$\omega_U(n_i)$		10	4	4	2	2	4							
$\omega_U(m)$		4	4	2	2	2	4							
$\omega_U(m-v)$		2	2	2	0	0	2							
$h_j\chi'_j(n_i)$		30	0	−12	−6	0	0							
$h_j\chi_j(T)$		3	0	−3	−3	0	0							
$h_j\chi_j(T')$		9	0	−3	−3	0	0							
$h_j\chi_j(R')$		6	0	−6	0	0	0							

(R = Raman, I = infrared, a = allowed, f = forbidden.)

3.7.5 The determination of the normal modes

Consider the normal mode q_k which belongs to an irreducible representation Γ_k. Three types of q_k exist depending upon the nature of Γ_k:

(i) Γ_k is non-degenerate having only one q_k, and q_k will be unique.

(ii) Γ_k is non-degenerate but has a number of normal modes q_k. For any unit cell group operation $\mathscr{U}_k$

$$\mathscr{U}_k \cdot q_k = \chi_k^*(U_k) \cdot q_k, \tag{3.27}$$

where $\chi_k(U_k)$ is the character of U_k in the irreducible representation $= +1$ or -1 for most unit cell groups. In this case there will be a number of combinations ($=$ number of q_k for Γ_k) that satisfy (3.27) and these must be mutually independent and orthogonal to each other. The choice of the various q_k will not however be unique in some cases and will depend upon the interatomic and intermolecular forces present.

(iii) Γ_k is g-fold degenerate and has a number of normal modes q_k. The same situation arises as in (ii) except that each mode will be g-fold degenerate.

A method must be found for expressing the normal modes in terms of the vectors of the coordinate system that describes the atomic and molecular displacements (the cartesian displacement coordinates). Diagrams can then be drawn to represent both the normal modes, and these can be modified to represent the normal coordinates Q_k if the arrows are drawn to represent displacements in a mass adjusted scale. The need for the mass adjusted scale arises because normal coordinates are such that they allow the kinetic and potential energies to be written down, for harmonic forces, as the sum of the squared terms:

$$\text{potential energy} = \tfrac{1}{2}\sum_k \lambda_k Q_k^2, \qquad (3.28)$$

where λ_k is a coefficient that represents the force constant ($\omega = (\lambda_k/m)^{\frac{1}{2}}$), and Q_k is considered to have absorbed the mass term,

$$\text{kinetic energy} = \tfrac{1}{2}\sum_k \frac{\partial Q_k^2}{\partial t} = \tfrac{1}{2}\sum_k \dot{Q}_k^2 \qquad (3.29)$$

since $v = \partial Q_k/\partial t$ and m is absorbed into Q_k. Q_k will therefore have to be expressed in terms of mass weighted cartesian displacement coordinates.

There will be $(1/m)(3n-6m)$ *internal coordinates* (these are displacement coordinates that describe the internal configuration of the molecule without regard to its position in space) corresponding to displacements of atoms in the molecules. For an isolated molecule of known symmetry it is possible to consider the molecular symmetry operations on the internal coordinates and hence obtain which internal coordinates are contained in various irreducible representations of the molecular point group [252]. The situation in the unit cell is rather more complicated because the molecules are not independent. There will be m different modes for each of the modes described by one set of different coordinates (the correlation field components of the internal modes), depending upon the m possible phase arrangements for those vibrations (i.e. all in phase, or wholly or partially out of phase). There will be $3m$ displacement vectors that describe the rotation of entire molecules and $3(m-1)$ displacement vectors that describe translation of entire molecules, and three displacement vectors describing the acoustic modes describing displacement of

Group theory applied to crystal lattice

the entire unit cell. The determination of the displacement coordinates that are contained in various irreducible representations of the unit cell group becomes a more complicated problem therefore than for the case of the isolated molecule. These displacement coordinates are required to draw diagrams for the normal modes and an often simpler way of determining the displacement coordinates is to consider the projection operator [253].

If the vectors of the coordinate system (the cartesian displacement coordinates) are represented by $f(r)$, then a projection operator $\mathscr{P}_k$ which projects the part of $f(r)$ which belongs to the irreducible representation Γ_k under consideration can be defined:

$$\mathscr{P}_k \cdot f(r) = \frac{g}{Z} \sum_x \chi_k^*(U_x) \cdot \mathscr{U}_x \cdot f(r). \qquad (3.30)$$

Balkanski and Teng [251] have used this technique for obtaining the normal modes of the various forms of potassium nitrate in terms of the cartesian displacement coordinates. This allowed diagrams of the displacement vectors that described the normal modes to be drawn. For the example of fig. 3.1 the cartesian displacement coordinates of the atoms x_{1-10}, y_{1-10}, and z_{1-10} can be taken as $f(r)$. There are six irreducible representations of the unit cell group allowing six projection operators $\mathscr{P}_{1-6}$ to be considered. These projection operators are then used to operate on the cartesian displacement coordinates in the different modes. The results of (3.30) can be achieved by considering the six different types of operator $\mathscr{U}_x$ and evaluating the operation

$$\sum_x \mathscr{U}_x \cdot f(r) \qquad (3.31)$$

for all $f(r)$, i.e. x_{1-10}, y_{1-10}, and z_{1-10}. For x_4,

$$\sum_x \mathscr{U}_x \cdot x_4 = (x_{4E} + x_{4C_2} + x_{4C_2} + x_{4i} + x_{4S_6} + x_{4\sigma_{d(\text{glide})}}), \qquad (3.32)$$

where x_{4U_x} represents $\mathscr{U}_x \cdot x_4$. The vectors x_{4U_x} can easily be drawn out since x_4 lies opposite $O8$ (fig. 3.1) and the effects of the symmetry operation $\mathscr{U}_x$ on 8 is given in table 3.4. The vectors x_{4U_x} are shown in fig. 3.2 and it can be seen that the resultant vector is zero. Thus

$$\sum_x \mathscr{U}_x \cdot x_4 = 0. \qquad (3.33)$$

Likewise it can be shown that

$$\sum_x \mathscr{U}_x \cdot x_1 = \sum_x \mathscr{U}_x \cdot x_2 = \sum_x \mathscr{U}_x \cdot x_3 = 0. \qquad (3.34)$$

Unit cell group analysis

For z_4 the vectors z_{4U_x} lie along O2 (fig. 3.1) and the effects of the symmetry operation $\mathcal{U}_x$ on 2 is given in table 3.4. Thus

$$\sum_x \mathcal{U}_x \cdot z_4 = 0, \qquad (3.35)$$

and likewise

$$\sum_x \mathcal{U}_x \cdot z_1 = \sum_x \mathcal{U}_x \cdot z_2 = \sum_x \mathcal{U}_x \cdot z_3 = 0. \qquad (3.36)$$

By considering the atoms 8, 9, 10, 5, 6, 7 and the effects of the symmetry operation $\mathcal{U}_x$ on them in table 3.4 it can be shown that

$$\sum_x \mathcal{U}_x \cdot z_{5-7, 8-10} = 0. \qquad (3.37)$$

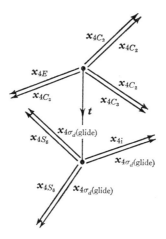

Fig. 3.2. Vectors of magnitude $|x_4|$ mapped out by operations $\mathcal{U}_x \cdot x_4$ (represented x_{4U_x}). x_4 is the cartesian x displacement coordinate of atom 4 (fig. 3.1).

Similarly

$$\sum_x \mathcal{U}_x \cdot y_{1-4, 5, 8} = 0, \qquad (3.38)$$

$$\sum_x \mathcal{U}_x \cdot y_{6, 7, 9, 10} = \text{finite}, \qquad (3.39)$$

and

$$\sum_x \mathcal{U}_x \cdot x_{5-10} = \text{finite}. \qquad (3.40)$$

In the case of the finite quantity in (3.39) and (3.40) it is in all cases proportional to the function f':

$$f' = 2(x_5 - x_8) - (x_6 + x_7 - x_9 - x_{10}) - (\sqrt{3})(y_6 - y_7 - y_9 + y_{10}). \qquad (3.41)$$

The projection operators $\mathcal{P}_{1-6}$ can now be used to evaluate the 30 normal modes and 30 normal coordinates. To obtain the normal coordinates

Group theory applied to crystal lattice

mass adjusted displacement coordinates are introduced by using the following masses:

m_1 = mass of atoms 1 and 2,

m_2 = mass of atoms 3 and 4,

m_3 = mass of atoms 5, 6, 7, 8, 9 and 10.

In order to evaluate (3.30), (3.32) to (3.41) are multiplied by

$$\frac{g}{Z}\sum_{x}\chi_k^*(U_x). \tag{3.42}$$

The normal modes belonging to the various irreducible representations Γ_k of the unit cell group can now be evaluated in terms of the displacement coordinates.

(i) $\Gamma_1 = A_{1g}$

(3.42) = 1, thus only one mode given by (3.41) results:

$$q_1(\nu_1) = \frac{Q_1(\nu_1)}{\sqrt{m_3}} = x_5 - x_8 - (x_6 + x_7 - x_9 - x_{10})$$
$$- (\sqrt{3})(y_6 - y_7 - y_9 + y_{10}). \tag{3.43}$$

This mode is an internal mode corresponding to symmetric stretching of the XY_3 groups (ν_1).

(ii) $\Gamma_2 = A_{2g}$

Γ_2 is non-degenerate and has 3 normal modes q_2. Two of these three modes will be external modes, one a translational mode and one a rotational mode, thus the three normal modes can be separated according to their vibration type. (3.42) inverts the direction of the vectors x_{U_x}, y_{U_x} and z_{U_x} when $\mathscr{U}_x = 3C_2$ and $3\sigma_d$ (glide). On evaluating (3.30) it is found that the internal mode, which corresponds to the deformation of the XY_3 groups (ν_2) is

$$\left.\begin{aligned}q_1^1(\nu_2) &= z_3 - z_4 - (z_5 + z_6 + z_7 - z_8 - z_9 - z_{10}), \\ Q_1^1(\nu_2) &= (\sqrt{m_2})(z_3 - z_4) - (\sqrt{m_3})(z_5 + z_6 + z_7 - z_8 - z_9 - z_{10})\end{aligned}\right\} \tag{3.44}$$

the external mode, which corresponds to the translation in the z direction is

$$\left.\begin{aligned}q_2^2(T') &= z_3 - z_4 + z_5 + z_6 + z_7 - z_8 - z_9 - z_{10}, \\ Q_2^2(T') &= (\sqrt{m_2})(z_3 - z_4) + (\sqrt{m_3})(z_5 + z_6 + z_7 - z_8 - z_9 - z_{10})\end{aligned}\right\} \tag{3.45}$$

Unit cell group analysis

and the rotational mode, which corresponds to rotation in the xy direction is

$$q_2^3(R') = \frac{Q_2^3(R')}{\sqrt{m_3}} = (y_5 - y_8) - (y_6 + y_7 - y_9 - y_{10})$$
$$-(\sqrt{3})(x_7 - x_{10} - x_6 + x_9). \quad (3.46)$$

(iii) $\Gamma_3 = E_g$

Γ_3 is 2-fold degenerate ($g = 2$) and has $4 \times 2 = 8$ normal modes q_3. 1×2 of these will be translational external modes, 1×2 of these will be rotational external modes, and 2×2 of these will be internal modes. In this case, the q_3 cannot be separated each into a different type of vibrational motion (translational, rotational and internal) as was done for q_2, and thus the choice of q_3 will not be unique. Eight mutually independent and orthogonal functions must be searched for that belong to Γ_3. (3.42) inverts, modifies, and eliminates the vectors x_{U_x}, y_{U_x} and z_{U_x} when $\mathcal{U}_x = 2C_3$ and $2S_6$, E and i, $3C_2$ and $3\sigma_d$ (glide). On evaluating (3.30) it is found that two mutually independent and orthogonal functions for the two translational external modes can be chosen if translations in the x and y directions are considered:

$$\left.\begin{aligned}
q_3^{1\prime}(T') &= x_3 - x_4 + x_5 - x_8 + x_7 - x_{10} + x_6 - x_9, \\
Q_3^{1\prime}(T') &= (\sqrt{m_2})(x_3 - x_4) + (\sqrt{m_3})(x_5 - x_8 + x_7 - x_{10} + x_6 - x_9), \\
q_3^{1\prime\prime}(T') &= y_3 - y_4 + y_5 - y_8 + y_7 - y_{10} + y_6 - y_9, \\
Q_3^{1\prime\prime}(T') &= (\sqrt{m_2})(y_3 - y_4) + (\sqrt{m_3})(y_5 - y_8 + y_7 - y_{10} + y_6 - y_9),
\end{aligned}\right\} \quad (3.47)$$

and two mutually independent and orthogonal functions for the two rotational external modes can be chosen if rotations about the x and y axes are considered:

$$\left.\begin{aligned}
q_3^{2\prime}(R') &= \frac{Q_3^{2\prime}(R')}{\sqrt{m_3}} = z_7 + z_9 - z_6 - z_{10}, \\
q_3^{2\prime\prime}(R') &= \frac{Q_3^{2\prime\prime}(R')}{\sqrt{m_3}} = (z_5 - z_8) - z_6 - z_7 + z_9 + z_{10},
\end{aligned}\right\} \quad (3.48)$$

and four mutually independent and orthogonal functions for the four internal modes can be chosen if the asymmetric XY_3 group motions (ν_3) and (ν_4) are considered with XY motion in the x and y directions:

$$\left.\begin{aligned}
q_3^{3\prime}(\nu_{3a}) &= (x_3 - x_4) - (x_5 + x_6 + x_7 - x_8 - x_9 - x_{10}), \\
Q_3^{3\prime}(\nu_{3a}) &= (\sqrt{m_2})(x_3 - x_4) - (\sqrt{m_3})(x_5 + x_6 + x_7 - x_8 - x_9 - x_{10}), \\
q_3^{3\prime\prime}(\nu_{3b}) &= (y_3 - y_4) - (y_5 + y_6 + y_7 - y_8 - y_9 - y_{10}), \\
Q_3^{3\prime\prime}(\nu_{3b}) &= (\sqrt{m_2})(y_3 - y_4) - (\sqrt{m_3})(y_5 + y_6 + y_7 - y_8 - y_9 - y_{10}),
\end{aligned}\right\} \quad (3.49)$$

Group theory applied to crystal lattice

and

$$\begin{aligned}
q_3^{4'}(\nu_{4a}) &= (x_3-x_4)-(x_6+x_7-x_9-x_{10}) \\
&\quad +(\sqrt{3})(y_6-y_7-y_9+y_{10}), \\
Q_3^{4'}(\nu_{4a}) &= (\sqrt{m_2})(x_3-x_4)-(\sqrt{m_3})[(x_6+x_7-x_9-x_{10}) \\
&\quad +(\sqrt{3})(y_6-y_7-y_9+y_{10})], \\
q_3^{4''}(\nu_{4b}) &= (y_3-y_4)-(y_5+y_6+y_7-y_8-y_9-y_{10}) \\
&\quad +(\sqrt{3})(x_6-x_7-x_9+x_{10}), \\
Q_3^{4''}(\nu_{4b}) &= (\sqrt{m_2})(y_3-y_4)-(\sqrt{m_3})[(y_5+y_6+y_7-y_8-y_9-y_{10}) \\
&\quad +(\sqrt{3})(x_6-x_7-x_9+x_{10})].
\end{aligned} \quad (3.50)$$

(iv) $\Gamma_4 = A_{1u}$

Γ_4 is non-degenerate and has 2 normal modes q_4, which can similarly be shown to be

$$q_4^1(\nu_1) = \frac{Q_4^1(\nu_1)}{\sqrt{m_3}} = (x_5+x_8)-(x_6+x_7+x_9+x_{10})$$
$$-(\sqrt{3})(y_6-y_7+y_9-y_{10}), \quad (3.51)$$

$$q_4^2(T') = \frac{Q_4^2(T')}{\sqrt{m_1}} = z_1-z_2. \quad (3.52)$$

(v) $\Gamma_5 = A_{2u}$

Γ_5 is non-degenerate and has 4 normal modes q_5, which can similarly be shown to be

$$\left.\begin{aligned}
q_5^1(\nu_2) &= (z_3+z_4)-z_5-z_6-z_7-z_8-z_9-z_{10}, \\
Q_5^1(\nu_2) &= (\sqrt{m_2})(z_3+z_4)-(\sqrt{m_3})(z_5+z_6+z_7+z_8+z_9+z_{10}),
\end{aligned}\right\} \quad (3.53)$$

$$\left.\begin{aligned}
q_5^2(T) &= z_1+z_2+z_3+z_4+z_5+z_6+z_7+z_8+z_9+z_{10}, \\
Q_5^2(T) &= (\sqrt{m_1})(z_1+z_2)+(\sqrt{m_2})(z_3+z_4) \\
&\quad +(\sqrt{m_3})(z_5+z_6+z_7+z_8+z_9+z_{10}),
\end{aligned}\right\} \quad (3.54)$$

$$\left.\begin{aligned}
q_5^3(T') &= z_1+z_2-z_3-z_4-z_5-z_6-z_7-z_8-z_9-z_{10}, \\
Q_5^3(T') &= (\sqrt{m_1})(z_1+z_2)-(\sqrt{m_2})(z_3+z_4) \\
&\quad -(\sqrt{m_3})(z_5+z_6+z_7+z_8+z_9+z_{10}),
\end{aligned}\right\} \quad (3.55)$$

$$q_5^4(R') = \frac{Q_5^4(R')}{\sqrt{m_3}} = (y_5+y_8)-(y_7+y_{10}+y_6+y_9)$$
$$-(\sqrt{3})(x_7+x_{10}-x_6-x_9). \quad (3.56)$$

Unit cell group analysis

(vi) $\Gamma_6 = E_u$

Γ_6 is 2-fold degenerate ($g = 2$) and has $6 \times 2 = 12$ normal modes q_6, which can similarly be shown to be

$$\left.\begin{aligned}
q_6^{1\prime}(\nu_{3a}) &= (x_3+x_4)-(x_5+x_6+x_7+x_8+x_9+x_{10}), \\
Q_6^{1\prime}(\nu_{3a}) &= (\sqrt{m_2})(x_3+x_4)-(\sqrt{m_3})(x_5+x_6+x_7+x_8+x_9+x_{10}) \\
q_6^{1\prime\prime}(\nu_{3b}) &= (y_3+y_4)-(y_5+y_6+y_7+y_8+y_9+y_{10}), \\
Q_6^{1\prime\prime}(\nu_{3b}) &= (\sqrt{m_2})(y_3+y_4)-(\sqrt{m_3})(y_5+y_6+y_7+y_8+y_9+y_{10}),
\end{aligned}\right\} \quad (3.57)$$

$$\left.\begin{aligned}
q_6^{2\prime}(\nu_{4a}) &= (x_3+x_4)-(x_6+x_7+x_9+x_{10}) \\
&\quad +(\sqrt{3})(y_6-y_7+y_9-y_{10}), \\
Q_6^{2\prime}(\nu_{4a}) &= (\sqrt{m_2})(x_3+x_4)-(\sqrt{m_3})[(x_6+x_7+x_9+x_{10}) \\
&\quad +(\sqrt{3})(y_6-y_7+y_9-y_{10})], \\
q_6^{2\prime\prime}(\nu_{4b}) &= (y_3+y_4)-(y_5+y_6+y_7+y_8+y_9+y_{10}) \\
&\quad +(\sqrt{3})(x_6-x_7+x_9-x_{10}), \\
Q_6^{2\prime\prime}(\nu_{4b}) &= (\sqrt{m_2})(y_3+y_4)-(\sqrt{m_3})[(y_5+y_6+y_7+y_8+y_9+y_{10}) \\
&\quad +(\sqrt{3})(x_6-x_7+x_9-x_{10})].
\end{aligned}\right\} \quad (3.58)$$

$$\left.\begin{aligned}
q_6^{3\prime}(T) &= x_1+x_2+x_3+x_4+x_5+x_6+x_7+x_8+x_9+x_{10}, \\
Q_6^{3\prime}(T) &= (\sqrt{m_1})(x_1+x_2)+(\sqrt{m_2})(x_3+x_4) \\
&\quad +(\sqrt{m_3})(x_5+x_6+x_7+x_8+x_9+x_{10}), \\
q_6^{3\prime\prime}(T) &= y_1+y_2+y_3+y_4+y_5+y_6+y_7+y_8+y_9+y_{10}, \\
Q_6^{3\prime\prime}(T) &= (\sqrt{m_1})(y_1+y_2)+(\sqrt{m_2})(y_3+y_4) \\
&\quad +(\sqrt{m_3})(y_5+y_6+y_7+y_8+y_9+y_{10}),
\end{aligned}\right\} \quad (3.59)$$

$$\left.\begin{aligned}
q_6^{4\prime}(T') &= \frac{Q_6^{4\prime}}{\sqrt{m_1}}(T') = x_1-x_2, \\
q_6^{4\prime\prime}(T') &= \frac{Q_6^{4\prime\prime}}{\sqrt{m_1}}(T') = y_1-y_2,
\end{aligned}\right\} \quad (3.60)$$

$$\left.\begin{aligned}
q_6^{5\prime}(T') &= (x_1+x_2)-(x_3+x_4+x_5+x_6+x_7+x_8+x_9+x_{10}), \\
Q_6^{5\prime}(T') &= (\sqrt{m_1})(x_1+x_2)-(\sqrt{m_2})(x_3+x_4) \\
&\quad -(\sqrt{m_3})(x_5+x_6+x_7+x_8+x_9+x_{10}), \\
Q_6^{5\prime\prime}(T') &= (y_1+y_2)-(y_3+y_4+y_5+y_6+y_7+y_8+y_9+y_{10}), \\
Q_6^{5\prime\prime}(T') &= (\sqrt{m_1})(y_1+y_2)-(\sqrt{m_2})(y_3+y_4) \\
&\quad -(\sqrt{m_3})(y_5+y_6+y_7+y_8+y_9+y_{10}),
\end{aligned}\right\} \quad (3.61)$$

Group theory applied to crystal lattice

$$q_6^{6'}(R') = \frac{Q_6^{6'}(R')}{\sqrt{m_3}} = z_5 + z_8 - z_6 - z_7 - z_9 - z_{10},$$
$$q_6^{6''}(R') = \frac{Q_6^{6''}(R')}{\sqrt{m_3}} = z_7 - z_6 - z_9 + z_{10}.$$
(3.61 a)

Diagrams representing the displacement coordinates in these modes are given in fig. 3.3. It has been assumed that no mixing occurs between the modes (the subject of mixing between modes will be discussed in the following chapter).

Factor group analysis

The unit cell group analysis discussed in this section has frequently been called 'factor group analysis', but it has recently been pointed out [300] that the term 'factor group analysis' is unsuitable since one never uses the coset elements of the factor group in such an analysis, but always the elements of the unit cell group.

3.8 SITE GROUP ANALYSIS

In this analysis, originally developed by Halford [237], the atoms and molecules, other than the one considered, are kept in their equilibrium position. The molecule under consideration is then considered to vibrate in an environment of fixed symmetry given by its site group. The symmetry of the site group will nearly always be lower than that of the molecular point group, and in general the site group will be a subgroup of the molecular point group.

In the previous section it was seen how when $K = 0$ the types of modes of the crystal could be calculated from an analysis of the unit cell of smallest volume. In this case the types of mode (which will lead to the infrared and Raman activities) should be analysable on the basis of the subgroups of the unit cell group, the site groups. The molecules of the unit cell are therefore considered separately in terms of their local symmetry determined by some site group Si^x. The number of internal modes belonging to each irreducible representation of the site group is evaluated by counting the total number of atoms invariant under each symmetry operation, $\mathscr{S}i_x^x$, represented by the elements of the site group Si_x^x, and using (3.21) with S_x^x replacing U_x and (3.19). The total number of internal modes will be the sum of the number obtained from the site group analysis for each non-equivalent type of molecule. This total will be $(1/m)(n_1')$ since

Site group analysis

the site group analysis takes no account of correlation field splitting. A site group analysis will therefore tell us which, if any, of the degenerate internal modes of the free molecule will become non-degenerate as a result of the lower local symmetry of the molecule in the crystal.

TABLE 3.6 *Symmetry correlations for D_{3d}^6 space group*

Unit cell (2ZXY$_3$ molecules) D_{3d}^6	Site (One XY$_3$ group) D_3	Free XY_3 group D_{3h}
$A_{1g}(\nu_1)$		$A_1'(\nu_1)$
$A_{2g}(\nu_2)$		$A_2'(\nu_2)$
$E_g(\nu_3, \nu_4)$	$A_1(\nu_1)$	
$A_{1u}(\nu_1)$	$A_2(\nu_2)$	$E'(\nu_3, \nu_4)$
$A_{2u}(\nu_2)$	$E(\nu_3, \nu_4)$	A_1''
$E_u(\nu_3, \nu_4)$		A_2''
		E''

Consider the example of fig. 3.1. The molecules 3, 5, 6, 7 and 4, 8, 9, 10 (CO_3^{2-} or NO_3^- for example) have molecular symmetry D_{3h}. In the crystal the site symmetry is lower than the molecular symmetry and is D_3. For the molecule the normal modes are ν_1, ν_2, ν_3, ν_4 [254], where ν_3 and ν_4 are degenerate modes, ν_{3a}, ν_{3b} and ν_{4a}, ν_{4b}. In the crystal the degenerate modes are also degenerate in the site group D_3. In the unit cell group however it was noticed that the two correlation field components of each of these degenerate modes belonged to different irreducible representations of the unit cell group. The correlations between the irreducible representations of these various groups can be expressed by a *correlation table* (table 3.6). General lists of correlation tables can be found in many books on group theory and molecular spectroscopy. There are, of course, many other site groups for the example of fig. 3.1 other than the molecular site group which is of interest for site group analysis. Various other points have site groups S_6, C_i, C_3, C_2, and C_1 [11]. Sodium iodate, unlike the example above, provides an example of a crystal for which the degenerate internal modes of the free molecule become non-degenerate as a result of the lower local symmetry of the molecule in the crystal, and the correlation table [198] is given in table 3.7.

Site group analysis provides only partial information about crystal

Group theory applied to crystal lattice

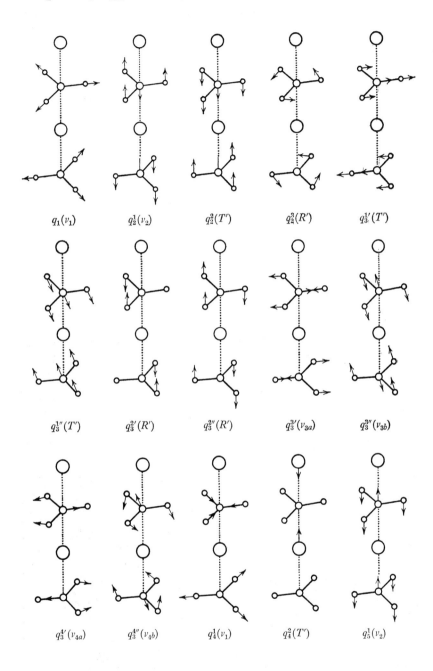

Unit cell group analysis

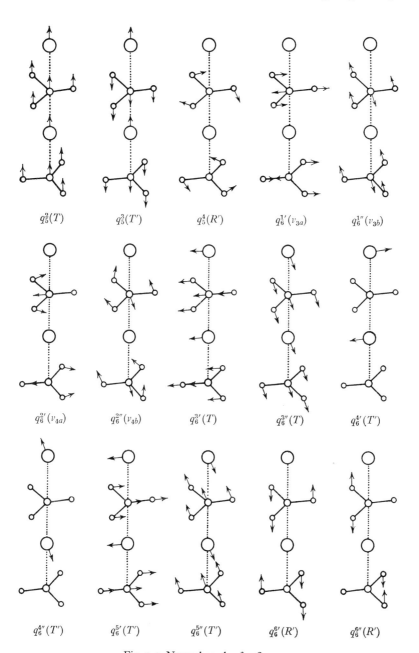

Fig. 3.3. Normal modes for fig. 3.1.

73

Group theory applied to crystal lattice

spectra since correlation field components and external modes are not considered. It is therefore better to consider the unit cell group analysis if at all possible. Of course if all the site groups are known the unit cell group can be derived. Halford [237] has produced tables which show the possible site groups for each of the 230 space groups. Adams and Newton [246, 247] have produced more detailed tables which allow the unit cell group to be derived from the site groups of each of the *atoms* present. Such a derivation of the unit cell group precludes any errors due to the choice of too large a unit cell, if only non-equivalent sites (or the minimum number of equivalent sites necessary to be consistent with the compound formula) are considered.

TABLE 3.7 *Symmetry correlations for D_{2h}^{16} space group*

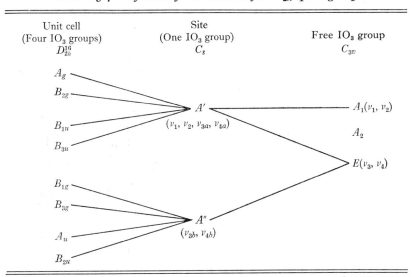

3.9 INFRARED AND RAMAN SELECTION RULES

Electromagnetic radiation has been seen to be composed of an electric and a magnetic field at right angles. The electric field (and the magnetic field) have been shown to be expressed by a sine wave, and so the electric field changes sign every half cycle. It is clear that such a field could transfer energy to any transition involving a dipole (for this dipole could be 'driven' by the electric field, and thus energy could be transferred). For any electronic or vibrational change, a transition between an initial state ψ_i, and a final state ψ_f will only occur if there is a dipole moment

Infrared and Raman selection rules

associated with the transition density $\psi_i \psi_f^*$. This is clearly the first requirement for an absorption process to occur. This requirement will now be explained in a little more detail. The basic problem is to consider the interaction that occurs between material particles and electromagnetic radiation. Quantum equations of motion of the electromagnetic field are found that are analogous to Maxwell's equations:

$$\nabla \times \boldsymbol{H} = \frac{1}{c}\left(\frac{\partial \boldsymbol{E}}{\partial t} + 4\pi \frac{\partial \boldsymbol{P}}{\partial t}\right), \tag{3.62}$$

$$\nabla \times \boldsymbol{E} = -\frac{1}{c}\frac{\partial \boldsymbol{H}}{\partial t}, \tag{3.63}$$

$$\nabla . \boldsymbol{H} = 0, \tag{3.64}$$

$$\nabla . \boldsymbol{E} = 0. \tag{3.65}$$

The interaction of the electromagnetic field with the material particles is then studied by quantum mechanical methods. For non-ionic solids the primary interaction is with the electrons and not with the nuclei, the vibrational motion of the nuclei being excited in a separate step by interaction of the electronic states with the vibrational states [255]. In ionic solids the primary interaction leads to an electric polarisation field that strongly interacts with the ions to give very strong absorption.

For any transition the transition probability per unit time, T, will be important. It can be shown [256] that T is given by the time independent expression

$$T = \frac{2g(\nu)}{\hbar}|H'_{fi}|^2, \tag{3.66}$$

where $g(\nu)$ is the density of states at the energy of the typical final state ψ_f (the concept of density of states has been discussed in §2.8), and H'_{fi} is the matrix element between ψ_f and ψ_i of the perturbation which induces the transition. It follows that the transition probability per unit time is not well defined unless the final states are relatively evenly distributed in energy. If $\boldsymbol{H}'$ represents the interaction of a radiation field, H'_{fi} is given by

$$H'_{fi} = \int \psi_f^* \boldsymbol{H}' \psi_i \, d\tau. \tag{3.67}$$

$\boldsymbol{H}'$ is describable by a vector potential $\boldsymbol{A}$ (defined by $\boldsymbol{H}' = \nabla \times \boldsymbol{A}$) with all the electrons in the system. $\boldsymbol{A}$ may be chosen to have a plane-wave form

$$\boldsymbol{A}(\boldsymbol{r}, t) = A_0 \boldsymbol{\xi} e^{-i(\boldsymbol{k} \cdot \boldsymbol{r} - \omega t)}, \tag{3.68}$$

where $\boldsymbol{\xi}$ is a unit vector in the direction of polarization of the wave. An

Group theory applied to crystal lattice

expression for the absorption probability can now be derived [257]. This expression will contain integrals which, when the wavelength of the radiation is many times greater than the linear dimensions of the wave functions that describe the motion of the particle (which is true in most cases), take the form

$$\frac{i}{\hbar}\int \psi_f^* p_\xi \psi_i \, d\tau, \qquad (3.69)$$

where p_ξ is the component of the particle momentum along the direction of polarization of the incident radiation (ξ). Transitions for which substitution of this integral into the absorption probability expression gives the absorption probability are called electric-dipole transitions, since only the matrix element of the electric-dipole moment, er, of the particle of charge e is involved. It follows, therefore, that for all electric-dipole transitions the integral must be non-zero in order that a transition may occur.

For a vibrational transition clearly the integral

$$\int \psi_f^* \mu_\xi \psi_i \, d\tau \qquad (3.70)$$

must be finite. Here μ_ξ is the component of the dipole moment of the vibrational transition along the direction of polarization of the incident radiation. This expression will now be used to deduce selection rules for vibrational transitions.

The above theory applies to an absorption process of the type which occurs in infrared studies. In Raman studies a scattering process is involved and another expression which can be used to deduce selection rules in such a process must be derived. Interaction can occur between electromagnetic radiation and the electrons in a crystal to give scattering because the electric field strength of the radiation induces a dipole moment **M** in the crystal. As a result of the wave nature of the radiation the induced dipole moment is varying. In elastic scattering the varying induced dipole moment can emit radiation identical in frequency to the incident radiation, but scattered over 360°. In inelastic scattering (observed in Raman studies) energy can be 'removed' from this induced dipole moment if a phonon can be produced that corresponds to a varying dipole moment during its motion. The scattered photon is scattered over 360° with phonons of different K at different angles (fig. 2.23), such inelastic scattering requires a polarizability change to be associated with the

Infrared and Raman selection rules

phonon. The 'amount', X, of varying dipole moment that can be 'removed', from the field induced dipole moment, is given by

$$|X| = \alpha_\xi \cdot |E|. \qquad (3.71)$$

Here E is the electric field strength of the incident radiation, and α_ξ the component of the polarizability tensor of the phonon, α, along the direction of polarization of the incident radiation. α_ξ has one component for an isotropic molecule but six components for a non-isotropic molecule:

$$\alpha_{xx}, \alpha_{yy}, \alpha_{zz}, \alpha_{xy}, \alpha_{zy}, \alpha_{yz}, \qquad (3.72)$$

which take account of unequal polarizability along the different principal axes of the molecule. Three equations can be written for X_A along the different principal axes of the crystal:

$$X_x = \alpha_{xx} E_x + \alpha_{xy} E_y + \alpha_{zz} E_z, \qquad (3.73)$$

$$X_y = \alpha_{yx} E_x + \alpha_{yy} E_y + \alpha_{yz} E_z, \qquad (3.74)$$

$$X_z = \alpha_{zx} E_x + \alpha_{zy} E_y + \alpha_{zz} E_z, \qquad (3.75)$$

where
$$\alpha_{xy} = \alpha_{yx}, \quad \alpha_{yz} = \alpha_{zy}, \quad \alpha_{zx} = \alpha_{xz}. \qquad (3.76)$$

Thus the integral
$$\int \psi_f^* \alpha_\xi \psi_i \, d\tau \qquad (3.77)$$

must be finite for a vibrational transition to be observed in the Raman spectrum.

3.9.1 Derivation of activities in unit cell and site group analysis

For any normal mode of the crystal the integrals (3.70) and (3.77) must be non-zero in order that the mode be infrared and/or Raman active.

When $K = 0$ the normal modes can be determined from a unit cell group analysis. The components of the dipole moment and the polarizability tensor can both be given by various representations, the characters, $\chi_k^{p_A}$, being given in the character tables of the unit cell group under consideration (table 3.5 and appendix). In order that the integrals (3.70) and (3.77) are non-zero they must be integrals of even functions. The wave functions ψ_i and ψ_f can be expressed in terms of the characters of the irreducible representation of the unit cell group that describes the particular vibrational mode, Γ_k^i and Γ_k^f, and therefore

$$\frac{1}{Z} \sum_x h_y \chi_{k_f}^*(U_x) \cdot \chi_k^{p_A}(U_x) \cdot \chi_{k_i}(U_x) = 1, \qquad (3.78)$$

Group theory applied to crystal lattice

if the integrals (3.70) and (3.77) are to be of even functions. It should be noted that the initial state is usually the ground state which is of course totally symmetric, and thus $\chi_{k_i}(U_x)$ can be omitted from (3.78). For such a case the components of the dipole moment and/or polarizability tensor must belong to the same irreducible representation of the unit cell group.

For $K = 0$, multiphonon processes can be determined in the same way. For a two-phonon process $\psi_i = \psi_{i_a}.\psi_{i_b}$, and $\psi_f = \psi_{f_a}.\psi_{f_b}$, where a and b refer to the two phonons. The characters of the representations of the higher vibrationally excited states can be calculated, for non-degenerate modes, by products of the characters of the ground state representations. Thus for the nth excited state

$$\chi_k^n(U_x) = (\chi_{k_i}(U_x))^n. \qquad (3.79)$$

For degenerate vibrations, however, the characters cannot be evaluated using (3.79) since the irreducible representations of the excited states involve several other irreducible representations [258]. These activity calculations are, however, of little value since most multiphonon processes occur for $K \neq 0$.

Activities of multiphonon modes for $K \neq 0$

For $K \neq 0$, the product of the number of classes of the star of K, and the group of K equals the number of classes of the space group, but, in general, the group of K is not identical with the factor group. The selection rules will now be determined by using (3.78) where $\chi_k(U_x)$ are replaced by space group characters $\chi_k(S_x)$. A number of authors [248, 259–68] have studied the problem, and Bradley and Cracknell [248] have discussed the full group and subgroup methods for solving the problem. In the special case of a completely arbitrary K, when the star of K contains H vectors, it can be shown [236] that the star of K will contain all unit cell group representations. The initial and final states will therefore involve a star of K, and the activity will depend upon the direct product of the star of K (which will, of course, always be an even function). Since the direct product will contain all the activity representations of the unit cell group, all combinations and overtones involving an arbitrary K (with an H-dimensional star of K) will be infrared and Raman active for all directions of polarized light or crystal orientation. For K values whose star contains fewer than H members, the selection rules will be stricter, though Wilson and Halford [236] have pointed out that, for a

Infrared and Raman selection rules

large crystal, these will constitute an asymptotically vanishing fraction of the total number of combinations. This is because H-dimensional stars will grow with large N as N^3, while the other types (stars less than H-dimensional) will grow as N^2 and N, and one dimensional stars ($K = 0$) will remain constant as given by the number of unit cell group representations.

Multiphonon bands will therefore fall into the following activity types.

(i) The *vast majority will be Raman and infrared active for all directions of polarized light or crystal orientations.*

(ii) There will be a number of intermediate modes whose selection rules will be stricter than for case (i) but much less strict than the rules for (iii).

(iii) There will be a small number of special cases for which $K = 0$, and these will have selection rules that can be predicted from unit cell group activity considerations.

3.10 EXAMPLES OF THE APPLICATION OF UNIT CELL GROUP ANALYSIS

Examples of the application of unit cell group analysis can be found in a number of reviews. Vedder and Hornig [4], Lawson [269] and Turner [270] give useful classified bibliographies for polyatomic crystals, which can be brought up to date by consulting specialist periodical reports [270]. Mitra and Gielisse [11] have reviewed the spectra of simple diatomic crystals, and Gilson and Hendra [15], Schrötter [271], Hester [272], Mathieu [9], and have discussed examples studied by Raman spectroscopy. Fince et al. [13], Mitra [12] and Wilkinson [16] present examples for a variety of different types of crystal. Griffith has considered the application to Raman studies of rock forming minerals [273], Ferraro [274] has reviewed the spectra of coordination compounds, and Bloor has given an extensive bibliography of far infrared spectroscopy [275].

The simplest type of example to consider is that of a crystal with one formula unit in the smallest volume unit cell. NH_4Cl [§5.5.3], UO_3, WO_3 [276], K_2UO_4, $CsUO_4$, $RbUO_4$ [277], I_2Cl_6 [278] and SiF_4 [279] provide some of the few examples of such cases. In these cases the site group symmetry of the molecule and the unit cell group symmetry will be identical. In a number of cases, however, it has been found [279] that the number of modes observed is larger than would be expected on the basis of a unit cell analysis. This initially disturbing result is due to splittings caused by various effects, and therefore in order to understand these effects the

Group theory applied to crystal lattice

mechanism of the interaction of radiation with a crystal will be examined in detail in later chapters.

The case of two formula units in the smallest volume unit cell provides a useful example of the use of unit cell group analysis, and the case of calcite and KNO_3 (phase I) has already been examined in detail. Other examples with various numbers of formula units in the smallest volume unit cell are listed in table 3.8. It should be remembered that, when single crystal studies are made, the orientation of the crystal can be selected. This allows the polarization of the Raman bands to be investigated and polarized infrared radiation to be used (the practical problems involved have been discussed [296]). A study of the irreducible representations of the unit cell group that contain the various components of the dipole moment and polarizability tensor in conjunction with such polarization studies will assist in normal mode assignment. Examples of polarization studies will be found in table 3.8.

The examples discussed so far have contained some ionic groupings. The unit cell group analysis of molecular crystals has been reviewed by Schnepp [118], Leroi [118 a], and Hadni [297], and table 3.9 lists some recent examples.

TABLE 3.8 *Examples of unit cell group analysis for crystals with various numbers of formula units (No_{Form}) in the smallest volume unit cell*

Formula	No_{Form}	Diagram of normal modes	Comments	Reference
$LiYF_4$	2	yes	Single crystal Raman	280
SnO_2	2	no	Polarized i.r. (single crystal)	281, 282
BaN_6	2	no	Single crystal Raman	283
$NaBF_4$	2	no	Powder i.r. + Raman	284
YVO_4	2	no	Single crystal i.r. and Raman	13, 285
LaF_3	2	no	Single crystal i.r. and Raman	286, 287
CeF_3, PrF_3, NdF_3	2	no	Polarized i.r. reflectance	287
$Cu(TA)_4Cl, Ni(TU)_4Cl_2$	2	no	Polarized i.r. + single crystal Raman	288
$PbCl_2$	4	yes	Single crystal Raman	289
$NaIO_3$	4	no	Powder i.r. + Raman	198
KH Malonate	4	yes	Single crystal Raman	290
Rare Earth Garnets	4	no	External modes all have same space group, one mode related to unit cell volume. Powder and pellet i.r.	291
Rare Earth stannates and titanate pyrochlores	2			292
Olivine type Silicates	4	no	Powder i.r.	293
Iron Pyrite	4	no	i.r. reflectance single crystal	294
$M(NO_3)_2$	8	no	M = Ca, Sr, Ba, Pb. Powder i.r. + Raman	295

N.B. TA = thioacetamide, TU = thiourea.

Examples of unit cell group analysis

TABLE 3.9 *Examples of unit cell group analysis of molecular crystals*

Coumpound(s)	Comments	Reference
	Inorganic solids	
CO_2, N_2O, N_2, CO	Raman spectra	298
O_2	Raman spectra	299
S	Far i.r.	300
Te	i.r. reflectance and absorption	301, 302
Se	i.r. reflectance and absorption	302
HCl	Raman spectra	303, 304, 305
H_2S, D_2S	Far i.r.	306
SO_2	Far i.r.	307
Br_2, CS_2, Cl_2	Raman spectra	308, 309
Br_2	Far i.r.	310
XeO_4	Raman spectra	311
OsO_4		
I_2O_5	Raman and i.r.	198
	Organic solids	
$(CH_3)_2SO_2$	Raman and i.r. single crystal	312
CH_2Cl_2, CD_2Cl_2	i.r.	313
$C_3N_3Cl_3$	Single crystal i.r. and Raman	314
CF_3CN	Raman and i.r.	315
Acetylene	Raman	316
Formic Acid	Raman and i.r.	317
Maleic Anhydride	Polarized single crystal i.r.	318
Succinic Anhydride	Polarized single crystal i.r.	318
Propanoic Acid	Polarized single crystal i.r.	319
Hexabromoethane	Raman and i.r. single crystal	320
Cyclohexane d_{12}	Raman and i.r. single crystal	321
Benzene	Torsional lattice modes	322
Pyridine	i.r.	324
Benzil	Raman single crystal	325
Naphthalene d_8	Raman and i.r. single crystal	326
1,2-dihydroxy cylcobutene-dione	Raman and i.r.	327
Tetracyano-quinodimethane and TCNQ-d_4	Polarized single crystal i.r.	328
π-benzenetricarbonyl chromium	Raman and i.r.	329
9,10-Anthraquinone	i.r. single crystal	330
1,4-Napthoquinone	i.r. single crystal	
Polyglycine	Raman spectra of KBr disc	331
2,2' paracyclophane	Powder and pellet i.r.	332
Adenine	Far i.r. and Raman	333
Uracil	Far i.r. and Raman	

Zbinden [201] has discussed the application of unit cell group analysis to polymeric systems, and Gilson and Hendra [15] have given examples of a number of Raman studies. The application of site group analysis to polymeric systems to find the local symmetry is of no use since the site

Group theory applied to crystal lattice

group contains point group operations, and a polymer chain is either two or three dimensional. Local symmetry can be described in terms of line groups (two dimensions) and plane groups or space groups (three dimensions giving the 'layer unit cell' [334]). A line group is a one-dimensional space group. A line group or plane group analysis will only be identical with the unit cell group analysis if there is only one layer or chain per primitive cell. Adams and Newton [246, 247] have explained how their tables for deriving the unit cell group can be adapted for polymeric systems. External vibrations of polymer systems are calculated assuming the polymer layers or chains to move as entire entities [198] but coupling between various internal and external modes can cause appreciable assignment difficulties (§2.8.5).

Theoretically the crystal structure of a crystal could be derived from a unit cell group analysis of a complete infrared and Raman study of all active normal modes. In the light of the complications to be discussed in later chapters, such as intensity differences, disorder, combination bands, size effects, such an analysis would be far from unambiguous. However infrared and Raman studies have been of value in correcting inexact crystallographic data. For example X-ray studies [335] suggest possible space groups D_2^6, C_{2v}^{11}, or D_{2h}^{19} for $Ba_2NaNb_5O_{15}$, and infrared and Raman studies show C_{2v}^{11} to be the correct space group [336]. Sometimes space groups that are very similar geometrically have sufficiently marked symmetry differences to cause more marked changes in the infrared and Raman spectra allowing corrections to be made. For example crystallographic studies of crystalline silicon tetrafluoride [337–8] have suggested a T_d^3 space group with one molecule per primitive unit cell. The infrared and Raman bands are consistent with molecules on sites of lower symmetry than T_d [279] the suggested sites being C_3 or C_{3v}. Possible space groups are C_3^1, C_3^4, C_{3v}^1, C_{3v}^2 and C_{3v}^5. $NbOCl_3$ provides another example. The crystal structure shows an infinite chain polymer with *trans* oxygen and *cis* chlorine bridges with a D_{4h}^{14} space group with four polymer chains in the unit cell [339]. Single crystal Raman studies [340] give better agreement with the very similar space group C_{4v}^4.

4 The interaction of radiation with a crystal

4.1 INTRODUCTION

In previous chapters it has been seen how certain lattice vibrations can be active as fundamentals in infrared and Raman studies. The mechanism of the interaction of the electromagnetic radiation with the crystal has so far only been discussed in a very general way. It is important to ask whether unit cell group analysis will always predict the correct number of active fundamentals or whether such predictions are not always reliable. The fact that the latter is true can easily be illustrated by considering a very simple example. Consider the case of a cubic infinite diatomic lattice (considered in chapter 2). Reference to table 3.1. shows that the point groups for cubic lattices are T, T_h, T_d, O, O_h. In chapter 2 it was seen that for a three-dimensional lattice there were three optical modes and these modes can easily be shown by a unit cell group analysis to belong to a triply degenerate representation F of the relevant point group (above) isomorphic to the unit cell group. In all cases reference to the point group character tables (appendix) shows that these triply degenerate vibrations should *all* be infrared (and for some point groups Raman) active. This is the case because the $K = 0$ transverse optical phonons, if the atoms are considered to be ions, give rise to components of the dipole moment of the vibrational transition μ_x and μ_z for phonon propagation in the y direction and atomic displacements in the xy and yz planes respectively. Also the longitudinal $K = 0$ optical phonon, if the atoms are considered to be ions, will give rise to a component of the dipole moment of the vibrational transition μ_y for phonon propagation in the y direction and atomic displacements in the y direction. The dispersion relation for a typical case (fig. 2.13) shows however that the transverse and longitudinal branches are not degenerate but that the transverse optical phonons are doubly degenerate and the longitudinal optical phonons occur at a higher frequency at $K = 0$. This longitudinal transverse splitting is thus not predicted by the unit cell analysis. Thus the absorption spectrum would be expected to contain two bands, one due to the transverse optical phonons and the other due to the longitudinal optical phonons. The absorption spectrum observed, however,

Interaction of radiation with a crystal

contains only one band due to the transverse optical phonons. It is easy to see why unit cell group analysis incorrectly predicts that the longitudinal optical phonon will be infrared active. In unit cell group analysis μ_ξ is the component of the dipole moment of the vibrational transition along the direction of polarization of the incident radiation for *each unit cell*. Since all unit cells move in phase at $K = 0$ this analysis applies to the

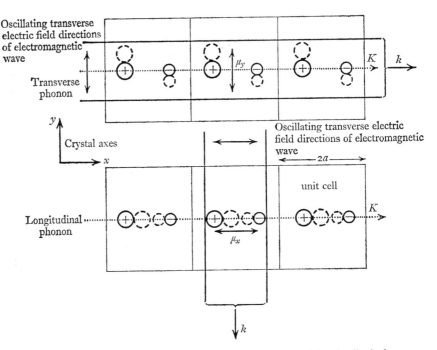

Fig. 4.1. Diagram illustrating the interaction of transverse and longitudinal phonons propagating in the direction K (with K direction represented by dotted line and ionic displacements by dashed circles) for a diatomic infinite ionic cubic lattice with $K = 0$ and the oscillating transverse electric field of an electromagnetic wave with $k = 0$ (with k direction represented by a solid line and electric field amplitude represented by two parallel solid lines).

whole crystal as far as the lattice vibration is concerned. However it *does not* apply to the whole crystal as far as μ_ξ is concerned. This is illustrated in fig. 4.1 which shows a transverse optical phonon and a longitudinal optical phonon extending over three unit cells of the crystal. Since the electric field of the electromagnetic radiation is a transverse field [341] it can couple with the μ_y of the transverse optical phonon illustrated for *all* the unit cells of the crystal but can only couple with the μ_x of the

Introduction

longitudinal optical phonon illustrated for *one* the unit cells of the crystal. Thus, for an infinite crystal, only the two transverse optical phonons would be expected to be infrared active.

It is more difficult to explain how the longitudinal transverse splitting arises and to do so one has to consider the macroscopic electrostatic field that arises from the displacement of ions during a lattice vibration. This leads to a discussion of the mechanism of interaction and also allows the questions below, which are essential to the correct analysis of a vibrational spectrum, to be answered.

(i) In infrared studies only absorption has been studied, but what effect do reflection and emission have?

(ii) It has seemed as if in the absorption process the photon has been destroyed and the phonon created, but might the photons interact with the phonons, and if so what effect might this have in infrared and Raman studies?

4.2 MODELS FOR STUDIES OF INTERACTION MECHANISM

It has been seen that energy can only be transferred from the photon to generate a phonon if the phonon is associated with a dipole moment change throughout the whole crystal. The displacement of ions will obviously lead to the generation of a dipole moment and this theoretical simplicity, together with the simplicity of structure of many ionic crystals, has meant that the mechanism of interaction for ionic crystals has been studied in detail. In §2.8, however, it was pointed out that the general results of a model based on Coulomb forces would be expected to be relevant to a covalent crystal with a finite q^*. Thus while a simple ionic model due to Huang [342] will be considered, the results will be of more general application. This theory allows longitudinal-transverse splitting to be explained, and an expression to be obtained for the dielectric constant in terms of the lattice vibration frequency for the transverse optical mode. The dielectric constant expression allows reflection to be considered as well as providing further experimental information for testing the theories of interaction mechanism.

4.3 LONGITUDINAL-TRANSVERSE SPLITTING AND THE DIELECTRIC CONSTANT FOR AN INFINITE CUBIC IONIC CRYSTAL

Consider a lattice in the continuum approximation ($K \approx 0$) with a dielectric constant ϵ. Fig. 4.2 shows the dispersion relation near $K = 0$

Interaction of radiation with a crystal

for some optical lattice vibration. The dispersion relation for an electromagnetic wave travelling through the lattice medium of refractive index, $n^l(\sqrt{\epsilon_\nu})$ will have a slope given by the phase velocity c/n^l, and fig. 4.2 shows that the slope is sufficiently great that intersection of the dispersion relation of the electromagnetic wave with that of the lattice vibration occurs at $K \approx 0$.

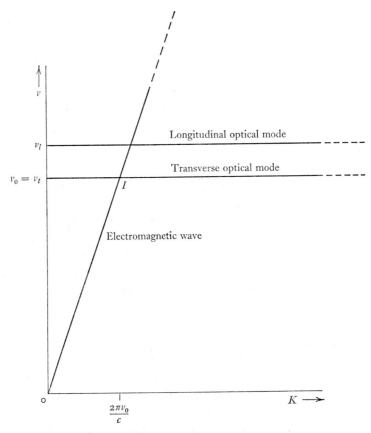

Fig. 4.2. Dispersion relation for transverse and longitudinal optical modes near $K = 0$ and an electromagnetic wave travelling through the crystal.

Huang described the motion of the ions in a diatomic infinite ionic crystal subject to an electric field $\boldsymbol{E}$, arising from the motion of the ionic charges excited by the electromagnetic wave, by the equation

$$A\mu'\ddot{\boldsymbol{u}} = b_{11}\boldsymbol{u} + b_{12}\boldsymbol{E}, \qquad (4.1)$$

where $\boldsymbol{u}$ is the displacement vector where μ' is the reduced mass for an

Longitudinal-transverse splitting

ion pair and A is the number of ion pairs in a unit volume (or the number of molecular pairs in a unit volume for an external mode of a covalent case). When microscopic models are considered later, $b_{11}\boldsymbol{u}$ will be seen to represent both the shorter range overlap forces and the longer range Coulomb attractions, and $b_{12}\boldsymbol{E}$ the longer range Coulomb attractions. $\boldsymbol{E}$ is the electric field that arises because of the movement of the positive ions against a rigid sub-lattice of negative ions during the lattice vibration. The polarization of the lattice, $\boldsymbol{P}$, will be made up of a part $b_{21}\boldsymbol{u}$ due to the displacement of the ions and another part $b_{22}\boldsymbol{E}$ induced by the electric field $\boldsymbol{E}$,

$$\boldsymbol{P} = b_{21}\boldsymbol{u} + b_{22}\boldsymbol{E}. \tag{4.2}$$

The polarization of the lattice will mean that the electric field of the lattice will be modified, and $\boldsymbol{E}$ in Maxwell equation (3.65) is replaced by $\boldsymbol{D}$,

$$\boldsymbol{D} = \epsilon\boldsymbol{E} = \boldsymbol{E} + 4\pi\boldsymbol{P}, \tag{4.3}$$

thus $\nabla \cdot \boldsymbol{D} = 0$ and (4.3) represents the application of Gauss' theorem to the crystal. It means that the total electric flux entering any volume element of the crystal is balanced by that leaving the element and must apply since the crystal is electrically neutral. In this treatment the lattice vibrations are assumed not to interact with electromagnetic radiation and the other three Maxwell equations are not considered. The conservation of energy requires that $b_{12} = b_{21}$, and the anharmonicity and higher order terms in the electric moment have been neglected in (4.1) and (4.2). Substituting (4.2) in the modified Maxwell equation (3.65):

$$\nabla \cdot \boldsymbol{E} = \frac{-4\pi b_{21}}{1 + 4\pi b_{22}} \nabla \cdot \boldsymbol{u}, \tag{4.4}$$

and the solenoidal and irrotational solutions of (4.4) are found to correspond to the transverse and longitudinal optical waves:

$$\nabla \cdot \boldsymbol{u}_t = 0 \text{ solenoidal}, \tag{4.5}$$

$$\nabla \times \boldsymbol{u}_l = 0 \text{ irrotational}, \tag{4.6}$$

and $$\boldsymbol{u} = \boldsymbol{u}_t + \boldsymbol{u}_l.$$

Clearly (4.4) can be rewritten with $\boldsymbol{u}_l$ replacing $\boldsymbol{u}$ (because of (4.5)), and an obvious solution for $\boldsymbol{E}$ is

$$\boldsymbol{E} = \frac{-4\pi b_{21}}{1 + 4\pi b_{22}} \cdot \boldsymbol{u}_l. \tag{4.7}$$

This unique solution shows that $\boldsymbol{E}$ must be irrotational. This means that the local macroscopic field $\boldsymbol{E}$ is zero for a transverse mode (this can be

Interaction of radiation with a crystal

seen by examining fig. 2.8(b)) and thus for a transverse mode (4.1) contains one term $b_{11}\mathbf{u}$. For the longitudinal mode, $\mathbf{E} \neq \mathbf{0}$ and (4.1) contains both terms and thus longitudinal modes would be expected to have a higher vibrational frequency, as illustrated in fig. 4.2.

The dielectric constant for any particular frequency can be deduced directly from (4.1) and (4.2) by considering periodic solutions (since the electric field of the electromagnetic wave excites ionic motion):

$$\mathbf{E} = \mathbf{E}_0 \cdot e^{-2\pi i \nu t}, \tag{4.8}$$

$$\mathbf{u} = \mathbf{u}_0 \cdot e^{-2\pi i \nu t}, \tag{4.9}$$

$$\mathbf{P} = \mathbf{P}_0 \cdot e^{-2\pi i \nu t}, \tag{4.9a}$$

and substituting for $\mathbf{E}$, $\mathbf{u}$, and $\mathbf{P}$ in (4.1) and (4.2):

$$-4\pi^2 \nu^2 A\mu' \mathbf{u} = b_{11}\mathbf{u} + b_{12}\mathbf{E}, \tag{4.10}$$

$$\mathbf{P} = b_{21}\mathbf{u} + b_{22}\mathbf{E}. \tag{4.11}$$

Eliminating $\mathbf{u}$, $\mathbf{P}$ and $\mathbf{E}$ can be related:

$$\mathbf{P} = \left(b_{22} + \frac{b_{12} \cdot b_{21}}{-b_{11} - 4\pi^2 \nu^2}\right) \mathbf{E}. \tag{4.12}$$

Introducing the dielectric constant from (4.3), it can be expressed as

$$\epsilon_\nu = 1 + 4\pi b_{22} + \frac{4\pi b_{12} b_{21}}{-b_{11} - 4\pi^2 \nu^2}, \tag{4.13}$$

which can be written:

$$\epsilon_\nu = \epsilon_\infty + \frac{\epsilon_0 - \epsilon_\infty}{1 - (\nu/\nu_0)^2}, \tag{4.14}$$

where

$$b_{11} = -4\pi^2 A\mu' \nu_0^2, \tag{4.15}$$

$$b_{22} = \frac{(\epsilon_\infty - 1)}{4\pi}, \tag{4.16}$$

$$b_{12}^2 = b_{21}^2 = A\pi\mu'(\epsilon_0 - \epsilon_\infty)\nu_0^2, \tag{4.17}$$

and $\nu_t = \nu_0$ (the infrared frequency), where

$$\nu_l = \left[\frac{\epsilon_0}{\epsilon_\infty}\right]^{\frac{1}{2}} \nu_t, \tag{4.18}$$

(4.18) being known as the Lyddane–Sachs–Teller relation [343]. ϵ_0 is the static dielectric constant measured when $\nu = 0$ and can be determined through capacitance measurements using lumped-circuit techniques at

Longitudinal-transverse splitting

audio- and radio-frequencies [165]. ϵ_∞ is the high-frequency dielectric constant and can be measured by measuring the wavelength dependence of the refractive index in the visible spectrum using the conventional method involving minimum deviation for large-angle prisms. From such measurements and with the use of the Lyddane–Sachs–Teller relation (4.13), the ν_l can be calculated from ν_t (obtained from infrared absorption studies).

The infrared frequency ν_0 can also be calculated using the static and high-frequency dielectric constants and the force constant (which may be calculated from compressibility data and other data (§2.9.2)) using the expression of Szigeti [344]:

$$\nu_0 = \left[\frac{g}{4\pi^2\mu}\left(\frac{\epsilon_\infty+2}{\epsilon_0+2}\right)\right]^{\frac{1}{2}}. \tag{4.19}$$

Table 4.1 lists experimental values for ν_0, ϵ_0, ϵ_∞, and calculated values for ν_0, and ν_l for various cubic crystals.

The theory so far has applied to an infinite diatomic cubic lattice with surroundings of tetrahedral symmetry, known as a diagonally cubic crystal (e.g. NaCl = O_h^5). The theory has been applied to the other special cases. Diagonally cubic crystals with any number of atoms in the unit cell have been considered [346], though the theory above can be applied if the crystal has only one infrared active vibration (e.g. CaF$_2$). Non-diagonally cubic crystals have also been considered [347, 348]. The theory applied to the special cases above can be considered for the general case [349] which applies to any crystal for which the adiabatic and harmonic approximations apply. For a lattice vibration, k ($k = 1, 2, 3$ are not considered as these correspond to acoustic modes) whose K points in the direction ξ, (4.18) can be written:

$$\prod_{k=4}^{3n}\left(\frac{\nu_k(\xi)}{\nu_{k0}}\right)^2 = \frac{\epsilon_0(\xi\xi)}{\epsilon_\infty(\xi\xi)}, \tag{4.20}$$

where (4.20) applies to any crystal of any symmetry. In the special case when ξ becomes one of the crystal axes (§2.4) and there are $n-1$ modes for each axis, (4.20) becomes

$$\prod_{n-1}\left(\frac{\nu_l(\xi)}{\nu_t(\xi)}\right)^2 = \frac{\epsilon_0(\xi\xi)}{\epsilon_\infty(\xi\xi)}, \tag{4.21}$$

which is identical to (4.18) when $n = 2$.

89

Interaction of radiation with a crystal

TABLE 4.1 *Lattice vibration parameters for body and face centred cubic lattices*

The data for ϵ_0, ϵ_∞, and ν_0(expt.) (except for that marked * which is taken from Kittel [24]) is taken from Lowndes and Martin [96]. The data for ν_0(calc.) is taken from compressibility data [345].

Crystal	ϵ_0	ϵ_∞	ν_0 (expt.)	ν_0 (calc.) from (4.19)	ν_l (calc.) from (4.18)
LiH	12.9*	3.6*	590*	–	1120
LiF	9.00	1.93	305	312	658
LiCl	11.86	2.75	203	227	422
LiBr	13.23	3.16	173	214	351.5
LiI	–	3.80	142	195	–
NaF	5.08	1.74	246.5	226	422
NaCl	5.90	2.33	164	170	270.5
NaBr	6.27	2.60	134	144	208.3
NaI	7.28	3.01	116	125	180.5
KF	5.50	1.85	194	–	345
KCl	4.84	2.17	142	139	212
KBr	4.90	2.36	114	113	164.2
KI	5.09	2.65	102	100	142.7
RbF	6.48	1.93	158	–	289.5
RbCl	4.89	2.18	116.5	114	174
RbFr	4.86	2.34	87.5	87	125.2
RbI	4.91	2.58	75.5	73	104
CsF	8.08	2.16	127	–	245.7
CsCl	6.95	2.63	99.5	95	161.8
CsBr	6.66	2.78	73.5	79	103.5
CsI	6.54	3.02	62.0	64	91.2
AgCl	11.14	3.92	105.5	–	177.5
AgBr	12.44	4.62	79.5	–	130.5
TlCl	32.6	4.76	63.0	–	164.7
TlBr	30.4	5.34	47.9	–	114.2

4.3.1 Microscopic models

So far the macroscopic dielectric properties have been calculated on the basis of a macroscopic model. These properties can also be calculated on the basis of microscopic models. Consider the charge on the ions of the diatomic infinite ionic crystal q^*. The polarization of the lattice, $\boldsymbol{P}$, is the dipole moment per unit volume:

$$\boldsymbol{P} = Aq^*\boldsymbol{u}. \tag{4.22}$$

The displacement force ($=$ mass $\times$ acceleration), $A\ddot{\boldsymbol{u}}$ will be given by the longer range Coulomb forces induced by the local electric field $\boldsymbol{E}_l$, $Aq^*\boldsymbol{E}_l$, and resisted by the shorter range overlap forces, $-Ag\boldsymbol{u}$ (the restoring force of Hooke's law):

$$A\mu'\ddot{\boldsymbol{u}} = Aq^*\boldsymbol{E}_l - Ag\boldsymbol{u}. \tag{4.23}$$

Longitudinal-transverse splitting

Lorentz has explained how a relation for the local field E_l can be obtained [350]. E_l is given by

$$E_l = E + E_1 + E_2. \qquad (4.24)$$

A sphere is considered about each atom such that the volume of the sphere is so large that the field from atoms outside the sphere, E_1, caused by the other atoms can be considered continuous. E_2 then represents the field from atoms within the sphere (this term can be considered, in the molecular case, to give rise to unit cell group splitting [350 a]). E_1 can be shown to have the value

$$E_1 = \frac{4\pi}{3} P. \qquad (4.25)$$

E_1 can deviate from the value given in (4.25) if overlap of the ionic charge clouds is considered [351, 165], and calculations of E_1 for spherically symmetrical charge clouds have been made [352, 353] which have led to a suggested 2 per cent deviation from (4.25) for some alkali halides.

E_2 will depend upon the crystal structure and for a diagonally cubic crystal will have the value zero (Born and Huang discuss the values for other structures [34]) Substituting the value of E_l (from (4.22), (4.24), and (4.25)) into (4.23):

$$A\mu' \ddot{u} = \left(\frac{4\pi}{3} A^2 q^{*2} - Ag\right) u + Aq^* E. \qquad (4.26)$$

(4.26) can now be compared with the macroscopic result (4.1) and the coefficients b expressed in terms of microscopic quantities, thus

$$b_{11} = -Ag + \frac{4\pi}{3} A^2 q^{*2}, \qquad (4.27)$$

where the first term represents the shorter range overlap forces and the second term the longer range Coulomb attractions, and

$$b_{12} = b_{21} = Aq^*. \qquad (4.28)$$

Comparing (4.22) with (4.2):

$$b_{22} = 0. \qquad (4.29)$$

The microscopic quantities can be equated with the macroscopic dielectric constants and frequencies by eliminating b from (4.27), (4.28), and (4.29) using (4.15) and (4.17):

$$4\pi^2 \nu_0^2 = \frac{-1}{A\mu'} \left(\frac{4\pi}{3} A^2 q^{*2}\right) - Ag, \qquad (4.30)$$

$$\epsilon_0 - \epsilon_\infty = \frac{Aq^{*2}}{\pi \mu' \nu_0^2}, \qquad (4.31)$$

$$\epsilon_\infty = 1. \qquad (4.32)$$

Interaction of radiation with a crystal

(4.14) can be modified to give ϵ_ν in terms of q^* by substituting (4.31) into (4.14):

$$\epsilon_\nu = \epsilon_\infty + \frac{Aq^{*2}}{\pi\mu'} \frac{1}{\nu_t^2 - \nu^2}. \qquad (4.33)$$

In a crystal where $q^* = 0$, $\epsilon_\nu = \epsilon_\infty$ and therefore by (4.18), $\nu_l = \nu_t$. Generally, at $K = 0$, $\nu_l > \nu_t$ and the difference will fall as q^* becomes smaller. Mitra [11, 12] has illustrated the effect of ionicity (size of q^*) on the dispersion curves of the zinc blende type of crystals. The dipole moment per ion pair caused by the displacement u in the direction ξ is denoted by μ_ξ (not to be confused with the reduced mass μ') where

$$\mu_\xi = P/A = q^*u, \qquad (4.34)$$

and for a particular lattice vibration k, u will be related to the normal coordinate of the vibration Q_k:

$$Q_k = \sqrt{\mu}\,|\mu|. \qquad (4.35)$$

Differentiating (4.34), after substituting Q_k for u:

$$\frac{\partial \mu_\xi}{\partial Q_k} = \frac{q^*}{\sqrt{\mu'}}, \qquad (4.36)$$

which allows q^* to be replaced by $\partial \mu_\xi / \partial Q_k$ in (4.33).

In the expression for the polarizability of the lattice (4.22), q^* was treated as if it were a point charge, but it will be remembered (§2.8) that q^* takes into account the polarization that occurs when the ions are displaced in a lattice vibration P_V. P_V makes a large contribution to ϵ_∞ and the neglect of P_V explains why the calculated value (4.32) is so different from the experimental values (table 4.1). The polarizability of the lattice p is modified to take P_V into account (see (3.71)):

$$\boldsymbol{P} = Aq^*\boldsymbol{u} + AP_V\boldsymbol{E}_l. \qquad (4.37)$$

The new $\boldsymbol{P}$ now takes ionic distortion into account. This can arise from the overlapping of an ion with its nearest neighbours [165, 354] and from the Coulomb-induced distortion due to the lack of uniformity, over an ion, of the field due to its near neighbours [165]. Substituting $\boldsymbol{E}_l = \boldsymbol{E} + \boldsymbol{E}_1$, with $\boldsymbol{E}_1 = \frac{4}{3}\boldsymbol{P}$ in (4.37):

$$\boldsymbol{P} = \frac{Aq^*}{1 - \frac{4}{3}\pi AP_V}\boldsymbol{u} + \frac{AP_V}{1 - \frac{4}{3}\pi AP_V}\boldsymbol{E}, \qquad (4.38)$$

Longitudinal-transverse splitting

and comparing (4.38) and (4.23) with the macroscopic results (4.1) and (4.2):

$$b_{12} = b_{21} = \frac{Aq^*}{1 - \frac{4}{3}\pi AP_V},\tag{4.39}$$

$$b_{22} = \frac{AP_V}{1 - \frac{4}{3}\pi AP_V},\tag{4.40}$$

$$b_{11} = -Ag + \frac{4}{3}\pi \frac{A^2 q^{*2}}{(1 - \frac{4}{3}\pi P_V)}.\tag{4.41}$$

Eliminating b from (4.39) and (4.40) using (4.15)–(4.17):

$$\epsilon_0 - \epsilon_\infty = \frac{Aq^{*2}}{\pi\mu v_0^2} \frac{(\epsilon_\infty + 2)^2}{9},\tag{4.42}$$

$$\epsilon_\infty = 1 + 4\pi AP_V/(1 - \tfrac{4}{3}AP_V),\tag{4.43}$$

and since [4, 34]

$$(\epsilon_\infty + 2)^2/9 = |E_l/E|,\tag{4.44}$$

(4.33) becomes (putting $\nu_0 = \nu_t$)

$$\epsilon_\nu = \epsilon_\infty + \left|\frac{E_l}{E}\right| \frac{Aq^{*2}}{\pi\mu'} \frac{1}{\nu_t^2 - \nu^2},\tag{4.45}$$

or replacing q^* using (4.36):

$$\epsilon_\nu = \epsilon_\infty + \left|\frac{E_l}{E}\right| \frac{A}{\pi} \left(\frac{\partial\mu_\xi}{\partial Q_k}\right)^2 \frac{1}{(\nu_{kt})^2 - \nu^2}.\tag{4.46}$$

4.3.2 Oriented gas model and static field shifting

The microscopic model discussed previously for a diatomic infinite ionic crystal where only three external optical modes occur can be applied to the internal optical modes of a molecular crystal in the *oriented gas model* [355, 356]. In the oriented gas model the intramolecular forces are assumed to be much stronger than the intermolecular forces and only the term $\sum_k V_k$ in the potential energy expression (1.1) is considered. Then if there is only one molecular species present and all the molecules are on equivalent sites, (4.46) can be used to give the general dielectric constant in terms of the various normal coordinates Q_k for internal vibrations of the *molecule* and the average frequency of *all* the *crystal* correlation field component frequencies $\bar{\nu}_{kt}$ (in the oriented gas model correlation field splitting will be zero, though in practice it is finite so the average must be used) [4, 357]:

$$\epsilon_\nu = \epsilon_\infty + \sum_k \left|\frac{E_l}{E}\right| \frac{A}{\pi} \left(\frac{\partial\mu_\xi}{\partial Q_k}\right)^2 \frac{1}{(\bar{\nu}_{kt})^2 - \nu^2}.\tag{4.47}$$

Interaction of radiation with a crystal

The value of $(\partial\mu_\xi/\partial Q_k)$ can be obtained from intensity measurements on the crystal infrared spectrum (to be discussed in the next chapter) or from measurements of the width of infrared reflection bands [357] (next section).

For the diatomic infinite ionic crystal g can be expressed in terms of an oscillator eigenfrequency ν_{0s}:

$$g = 4\pi^2 \mu' \nu_{0s}^2. \qquad (4.48)$$

This frequency, which corresponds to the molecular gas phase vibrational frequency when the diatomic ionic crystal model is applied to a molecular crystal in the oriented gas model, can be compared with $\bar{\nu}_t$. Substituting for b_{11} in (4.41) from (4.15) in

$$-4\pi^2 A \mu' \bar{\nu}_t^2 = -Ag + \tfrac{4}{3}\pi A q^* b_{21}, \qquad (4.49)$$

where b_{21} has been substituted from (4.39). Then substituting the values of g from (4.48) and b_{21} from (4.17):

$$\bar{\nu}_t^2 = \nu_{0s}^2 - \frac{(\epsilon_\infty+2)}{9}\frac{A}{\pi}\left(\frac{\partial\mu_\xi}{\partial Q_k}\right)^2. \qquad (4.50)$$

The second term on the right hand side of (4.50) thus represents the static field shifting of the gas phase molecular vibration frequency due to the term $\sum_k V_k$ in the potential energy expression of the crystal. Since the value of $(\partial\mu_\xi/\partial Q_k)^2$ is proportional to the solid state absorption intensity (§5.8) and will therefore be greater for fundamentals than for combinations or overtones, Haas and Ketelaar [354] pointed out that the static field shifting would be much greater for fundamentals (though $K \neq 0$ combinations and overtones complicate such considerations, and in addition such a conclusion is now considered erroneous [357 a]).

4.4 REFLECTION FOR AN INFINITE IONIC CRYSTAL

In general an electromagnetic wave falling on a crystal may be partly absorbed, and partly reflected, the remainder being transmitted. In addition, if the crystal temperature is not the same as the detector temperature, there will be emission. Finally depending upon the sample dimensions there may be elastic scattering (dependent upon particle size) and interference (dependent upon sample thickness). Reflection, absorption, and emission (for all possible multiple paths) are related through Kirchoff's law:

$$T(\nu) = 1 - A(\nu) - R(\nu) - \kappa(\nu), \qquad (4.51)$$

Reflection for infinite ionic crystal

where T = transmission, $A(\nu)$ = absorption,

$R(\nu)$ = crystal reflection coefficient = $I_{\text{reflected}}/I_{\text{incident}}$,

$\kappa(\nu)$ = emission coefficient = $\dfrac{\text{emission of crystal}}{\text{emission of black body}}$ at temp T,

and its determination can readily yield the absorption coefficient [12]. Reflection can be understood by considering the simple theory introduced in the previous section. The dielectric constant at a particular frequency (ϵ_ν) was given in (4.14), and it can be seen that when $\nu = \nu_0$, ϵ_{ν_0} becomes infinite. When ν is further increased by an infinitesimal amount, ν/ν_0 will have an infinitesimally small value making $\epsilon_\nu = -\infty$, and further increases in ν will make ϵ_ν finite and negative until ϵ_ν becomes zero. It can be seen from (4.14) that $\epsilon_\nu = 0$ when $\nu = \nu_l$ (given by (4.18)). The variation in ϵ and n with ν are given in figs. 4.3 and 4.4.

It should be noted that when ϵ is negative, n is imaginary though not zero (dotted line in fig. 4.4). Electromagnetic waves with frequencies that correspond to imaginary n values cannot therefore propogate through the crystal, and 100 per cent reflection occurs. In general the value of R for normal incidence can be obtained from n for a sample in air by using the Fresnel formula:

$$R = \left|\frac{n-1}{n+1}\right|^2, \qquad (4.52)$$

and fig. 4.5 shows the variation of R with ν derived from the values of n given in fig. 4.4.

The width of the reflection band can be seen to be $\nu_l - \nu_t$. Using the microscopic model of a diatomic infinite ionic crystal ν_l can be evaluated in terms of ν_t by using (4.46) or (4.47) and substituting $\epsilon_{\nu_l} = 0$ (fig. 4.2):

$$\nu_l^2 = \nu_t^2 + \frac{1}{\epsilon_\infty}\left|\frac{E_l}{E}\right|\frac{A}{\pi}\left(\frac{\partial \mu_\xi}{\partial Q_k}\right)^2. \qquad (4.53)$$

(4.53) is sometimes known as the Haas–Hornig equation [357]. Hence $(\partial \mu_\xi/\partial Q_k)$ can be obtained from the width of the reflection band $\nu_l - \nu_t$ using (4.53), and Haas and Hornig [357] give examples of such calculations from reflection spectra.

4.5 PHONON–PHOTON INTERACTION AND THE POLARITON

So far it has been assumed that there is no interaction between the electromagnetic radiation and the lattice vibrations. Interaction does however occur and the oscillations of the combined electromagnetic field

Interaction of radiation with a crystal

and lattice system must be analysed. In fact the interaction does not operate instantaneously but propagates with the velocity of light and the three other Maxwell equations (3.62), (3.63), and (3.64) must be considered [34]. The introduction of these three other Maxwell equations

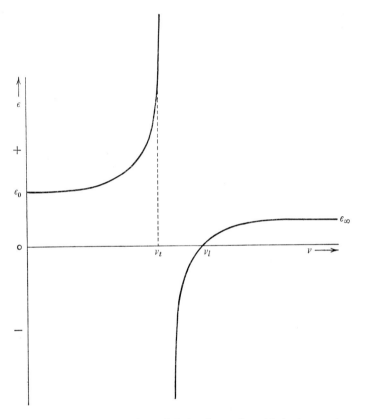

Fig. 4.3. (ϵ, ν) curves for an infinite diagonally cubic ionic crystal with no phonon–photon interaction.

has the effect of mixing the transverse lattice vibrations with the electromagnetic radiation. This mixing, known as *retardation* gives rise to a coupled excitation that is neither a pure phonon nor a pure photon but an excitation containing a variable amount of radiative and mechanical energy. The coupled excitation is known as a *polariton* and may be defined [358, 359] as polarization field 'particles' analagous to photons, where the polarization field is due to the long range polarization of the

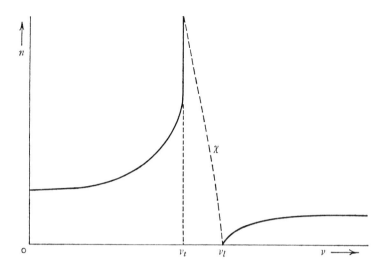

Fig. 4.4. (n, ν) curves for an infinite diagonally cubic ionic crystal with no phonon–photon interaction.

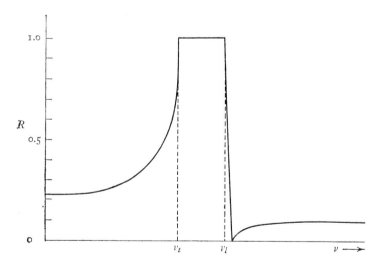

Fig. 4.5. Variation of the crystal reflection coefficient with frequency for an infinite diagonally cubic crystal with no phonon–photon interaction.

Interaction of radiation with a crystal

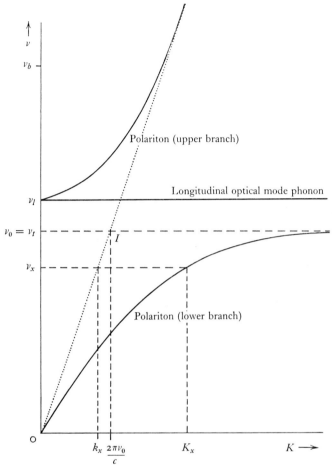

Fig. 4.6. Dispersion relation for longitudinal optical mode phonon and polaritons near $K = 0$.

crystal by the ionic motion. Polaritons arising from other polarization fields will be discussed in a later chapter.

The interaction discussed above means that fig. 4.2 showing the dispersion relation for the transverse optical mode and the electromagnetic wave must be modified at I where the dispersion relations cross. Elementary quantum mechanics shows that when two states of the same symmetry are coupled, they repel each other and cannot cross (the noncrossing rule) and thus the dispersion relations of fig. 4.2 in the region of I become modified accordingly (fig. 4.6). Fig. 4.6 now represents the

Phonon–photon interaction, the polariton

dispersion relation for the polaritons and the longitudinal optical mode phonon (which cannot, of course, mix with the electromagnetic wave), but different regions of the dispersion relation for the polaritons will

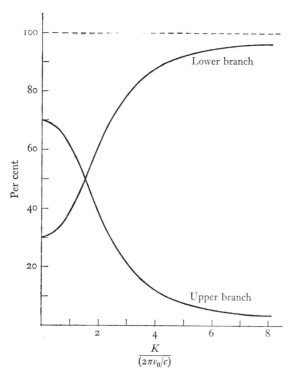

Fig. 4.7. Percentage mechanical energy of the polaritons calculated from the energy density [342]. The energy density is given by the expression:

$$\text{energy density} = \tfrac{1}{2}A\mu'(\dot{\boldsymbol{u}}^2+4\pi^2\nu_0^2\boldsymbol{u}^2)+\frac{1}{8\pi}(\epsilon_\infty E'^2+H^2)$$

derived by Born and Huang [34], where the first term represents the mechanical energy and the second term the radiative energy. The diagram is reproduced by permission of the Royal Society.

represent polaritons that are more or less phonon-like (have more or less mechanical energy). Fig. 4.7 shows the varying percentage of mechanical energy, which shows that polaritons of the upper branch are generally photon-like while those of the lower branch are generally phonon-like.

Interaction of radiation with a crystal

4.6 ABSORPTION, REFLECTION AND THE NEED FOR ANHARMONIC FORCES

The introduction of mixing in the previous section means that the absorption process can no longer be considered a process where the photon is destroyed and the phonon created. Consider photons travelling in free space approaching a crystal. On reaching the crystal surface a certain proportion of the photons will be reflected according to R, and a certain proportion $1 - R$ will enter the crystal. Consider the $(1 - R) \times 100$ per cent of the photons that enter the crystal with frequencies in the range 0 to ν_b (fig. 4.6). For such frequencies phonon (transverse optical)–photon interaction has become finite and thus on entering the crystal the photons become polaritons (of the upper or lower branch) where a photon of frequency ν_x and wave vector k_x gives rise to a polariton of frequency ν_x and wave vector K_x (fig. 4.6). When the photon frequency is ν_0 a very phonon-like polariton is formed, but under these conditions $R = 1$ and all the photons are reflected. This result is in conflict with experiment since in many cases finite transmission of electromagnetic radiation occurs under these conditions and the model of an infinite ionic crystal with harmonic forces must therefore be modified.

Anharmonic forces are repulsive forces whose presence in addition to harmonic attractive forces cause non-linearity of the cohesive forces of a solid. If the central frequency (2.23) ν_{ctr} is compared with ν_d the frequency calculated for harmonic forces (from lattice cohesive energy data for example [360, 361]) then a quantity $F(A)$ can be evaluated:

$$F(A) = \nu_{\text{ctr}}/\nu_d, \qquad (4.54)$$

where $F(A)$ is the factor of anharmonicity [210, 362]. For all solids except those with the argon configuration such as KCl, $F(A)$ deviates from unity indicating the presence of anharmonic forces, where the extent of anharmonic forces is represented by $1 - F(A)$.

Two types of lattice anharmonicity can be present [362]. In the 'hard force' case $F(A) > 1$, and the repulsive force arises from mutual deformation of the electron clouds of adjacent atoms without overlap, and in the 'soft force' case $F(A) < 1$, and the repulsive force arises from mutual deformation of the electron clouds of adjacent atoms by overlap. $F(A)$ has been shown to be a function of the reduced mass [362]

$$F(A) = X_m^{1/p}, \qquad (4.55)$$

Need for anharmonic forces; absorption, reflection

where p is usually equal to 9 and

$$X_m = \frac{\mu'}{31 \times 10^{-24}}. \tag{4.56}$$

$F(A)$ can also be evaluated from hardness and compressibility measurements [360].

In addition to the complication caused by the presence of anharmonic forces in most solids, real crystals are also not linear dielectrically as has been assumed in (4.1). Both these effects can be introduced into the preceding theory by introducing a frequency independent damping factor, γ. The damping factor introduces an additional repulsive force $-A\mu'\gamma\dot{u}$ into both the macroscopic and microscopic equations (4.1) and (4.23). This leads to a new expression for the dielectric constant:

$$\epsilon_\nu = \epsilon_\infty + \frac{\epsilon_0 - \epsilon_\infty}{1 - (\nu/\nu_0)^2 - i(\nu/\nu_0)\gamma}, \tag{4.57}$$

or

$$\epsilon_\nu = \epsilon_\infty + \left|\frac{E_l}{E}\right| \frac{A}{\pi}\left(\frac{\partial \mu_\xi}{\partial Q_k}\right)^2 \frac{1}{\nu_0^2 - \nu^2 - i\nu\gamma}. \tag{4.58}$$

Now when $\nu = \nu_0$, ϵ_{ν_0} is no longer infinite but has a large finite negative value whose size increases with increasing $(\partial \mu_\xi/\partial Q_k)^2$ and falls as ν is further increased from ν_0 becoming zero when $\nu \approx \nu_l$. n will no longer rise to infinity at $\nu = \nu_t$ and therefore R will no longer be 1 between ν_t and ν_l (fig. 3.4). In fact the width of the reflection band will be slightly greater than $\nu_l - \nu_t$ since $\epsilon_{\nu_l} \neq 0$ but

$$\epsilon_{\nu_l} = \epsilon_\infty - \epsilon_\infty \frac{(\epsilon_0 - \epsilon_\infty)}{(\epsilon_0 - \epsilon_\infty + i\gamma\epsilon_0)}, \tag{4.59}$$

and the error that this will cause in the use of (4.53) for the determination of $(\partial \mu_\xi/\partial Q_k)$ should be borne in mind. The value of R between ν_t and ν_l will depend upon the value of γ and $(\partial \mu_\xi/\partial Q_k)^2$ (which is proportional to the solid state absorption intensity). Thus for crystals with weak absorptions, R, will be much less than 1, while for strong absorptions, R will be large but, as long as $\gamma \neq 0$, R will *always* be less than 1. The frequency that corresponds to a maximum value of R is called the *Restrastrahl frequency*, ν_m, given by [363]:

$$\nu_m = \nu_t\left[1 + \frac{\epsilon_0 - \epsilon_\infty}{6\epsilon_\infty - 4}\right]. \tag{4.60}$$

Now that it has been seen that $R < 1$ for real crystals it is possible to

Interaction of radiation with a crystal

consider further the mechanism for absorption. We return to the $(1-R) \times 100$ per cent of the photons with the energy of a transverse optical phonon that can now (as $R < 1$) enter the crystal and in doing so become polaritons.

The polaritons now travel on through the crystal as a mixed phonon–photon eigenstate, which has a rapid exchange of energy between the phonon and the photon. The polaritons will travel on for ever in an infinite crystal and so the process represents an absorption, but in the real case of a finite crystal no absorption occurs. This is because the polaritons on reaching the other side of the crystal transfer all their phonon energy to photon energy and leave the crystal as a photon that has the same energy as the incident photon (with R per cent of the photons being reflected on leaving the other side of the crystal). In real crystals the presence of anharmonic terms and a non-linear dielectric nature represented by γ cause phonons to couple and no longer be mutually independent lattice vibrations. Since lower branch polaritons, for $\nu \approx \nu_t$, contain a high percentage of phonon energy (Fig. 4.7) even a small amount of phonon–phonon interaction (involving a multitude of different energy phonons) can lead to an appreciable reduction of the polariton energy flux leading to absorption of electromagnetic radiation by the crystal. It is important to realize that the energy absorbed by a crystal does not lie in the phonon modes to which the light was directly coupled, but instead lies in the phonon modes to which the polaritons can make transitions. Hopfield [358] points out that this qualitative difference can produce large changes in the interpretation of optical phenomena in crystals and that in order for the theory to be valid, it is necessary that the rate of exchange of energy between the phonon and photon in the polariton mixed eigenstate is fast compared to the transition probability for phonon–phonon interaction. If it were not the polariton would not be formed, and phonon–phonon interaction would occur first leading to a combination band rather than a fundamental. Combination bands therefore become more favoured at higher temperatures because the presence of more thermal phonons in the crystal leads to increased phonon–phonon interaction.

In an absorbing medium the dielectric constant is a complex quantity, and may be represented:

$$\epsilon_\nu = \epsilon'_\nu - i\epsilon''_\nu, \tag{4.61}$$

where the real and imaginary parts of the dielectric constant are related to one another in such a way that if ϵ''_ν is known for all frequencies then

Need for anharmonic forces; absorption, reflection

ϵ_v'' can be calculated for any particular frequency by using the Kramers–Kronig relations [3].

The refractive index, which is also complex, may be represented:

$$n = n' + i\chi, \tag{4.62}$$

where χ is the extinction coefficient for absorption. (4.60) and (4.61) can thus be used to replace ϵ_v and n in the preceding equations. Thus in fig. 4.4 the refractive index between ν_t and ν_l is represented by χ. χ is sometimes also represented by the symbol k [364], and is related to the absorption coefficient (defined in terms of the attenuation of the intensity, I, of a plane wave propagating in the material, $I = I_0 e^{-A'x}$ (see §5.8)) by

$$A' = 4\pi \frac{\nu_0}{c} \chi. \tag{4.63a}$$

Finally R is also complex in an absorbing medium, and may be represented [365]:

$$R = \left| \frac{n' - 1 + i\chi}{n' + 1 + i\chi} e^{i\delta} \right|^2, \tag{4.63b}$$

where δ is the phase difference between the incident and reflected rays. Thus

$$R = \frac{(n'-1)^2 + \chi^2}{(n'+1)^2 + \chi^2}. \tag{4.63c}$$

n' and χ will be given from (4.63 b) after separation into real and imaginary parts:

$$\left. \begin{array}{l} n' = \dfrac{1 - R}{1 + R - 2R^{\frac{1}{2}} \cos \delta}, \\[6pt] \chi = \dfrac{-2R^{\frac{1}{2}} \sin \delta}{1 + R - 2R^{\frac{1}{2}} \cos \delta}. \end{array} \right\} \tag{4.63d}$$

Thus n' and χ can be obtained from studies of the reflection spectrum. The problem is that δ is not readily measured in infrared studies. For isotropic solids n and χ may be obtained from values of R at two widely separated angles of incidence [365] (R depends upon the angle of incidence), but this method is unsuitable for anisotropic crystals where χ depends upon the angle of incidence. δ can be obtained by using the Kramers–Kronig dispersion relations [12, 3, 366, 367]. The method of obtaining the optical constants from the reflection spectrum is particularly useful in cases of samples with very intense absorption that makes transmission measurements very difficult.

Mitra [12] has described a damped oscillator model for the analysis of optical constants from the reflection spectrum of ionic solids (where R is large).

Interaction of radiation with a crystal

A quantum mechanical treatment of the vibrational properties of crystals [369–72, 12] leads to similar equations to those derived above classically, though the damping factor can be shown to be both frequency and temperature dependent. Thus γ is sometimes [12] replaced by a new factor (γ'/ν_0). Agreement between the two treatments may be less close in regions away from $\nu = \nu_0$ [12].

4.7 THE RAMAN SCATTERING MECHANISM

In the Raman scattering process monochromatic electromagnetic radiation of a frequency ν is inelastically scattered by the crystal. The frequency ν is chosen so that it is appreciably different from any ν_t of the crystal so that the amount of reflection and absorption is negligible, and thus ν is usually taken in the visible region of the spectrum where $\nu \gg \nu_t$. The mechanism of the process has been discussed in a general way in §2.8 and §3.9, and having now discussed the detailed mechanism of the absorption process, the detailed mechanism of the Raman process can be understood.

When the incident electromagnetic radiation reaches the crystal surface it enters the crystal and travels through the crystal. It is important at this stage to examine the exact nature of the electromagnetic radiation when it is travelling through the crystal. It was said above that $\nu \gg \nu_t$, and thus outside the range 0 to ν_b, meaning that the photon does travel as a photon through the crystal, but it should be remembered that excitations other than phonons can be generated by absorption of electromagnetic radiation. These other excitations will mix with the electromagnetic radiation and another polariton situation will arise and it will be necessary to consider the electromagnetic radiation travelling through the crystal as a polariton in such cases.

It was previously explained how the electric field strength of the electromagnetic radiation (or polariton) travelling through the crystal induces a varying dipole moment in the crystal. This varying dipole moment arises mainly from the electronic polarizability (the ionic polarizability being ignored except when ν is in the infrared region [372a]). If a classical model is considered then this electronic polarizability will give rise to an electron–hole pair under the influence of the electric field $\boldsymbol{E}$ of the electromagnetic radiation (given by (4.8)). The interaction, H_{ER}, can be expressed [5] either in terms of the position $\boldsymbol{x}$ of the electrons.

$$H_{ER} = -e\boldsymbol{E}.\boldsymbol{x}, \qquad (4.64)$$

The Raman scattering mechanism

where unfortunately the matrix elements of x have complicated properties or in terms of the vector potential A (3.68) and the momentum of the electrons p

$$H_{ER} = -\frac{eA \cdot p}{mc}. \tag{4.65}$$

Considering Coulomb attraction between the electron and the hole acting as a harmonic restoring force, the varying dipole moment, M, arising from the electronic polarizability can be written:

$$M = \frac{E_0 e^2}{4\pi^2 \mu} \frac{1}{\nu_0'^2 - \nu^2} e^{-2\pi i \nu t}, \tag{4.66}$$

where μ is the reduced mass and ν_0' is the natural vibration frequency of the electron–hole pair. Let the incident electromagnetic radiation, clothed by its associated electron–hole pair, have travelled a distance x through the crystal after a time t. At time t two possible instantaneous processes may occur.

(i) The dipole M may emit electromagnetic radiation of frequency ν_f (clothed by an electron–hole pair because of the electronic polarizability), where $\nu_f = \nu$, uniformly around its axis so that the amplitude of M is zero. If the incident radiation is vertically polarized the emitted radiation from M will have a maximum density at the equator and no intensity in the polar directions. This process corresponds to an elastic scattering event at t.

(ii) The electron of the electron–hole pair associated with the incident radiation may interact with an excitation (only phonons and polaritions will be considered in this chapter) if the excited state of the crystal corresponding to this excitation has a different electronic polarizability. The dipole that may emit radiation will now no longer be M but a different varying dipole moment M', where

$$|M'| = |M| \pm |X| \tag{4.67}$$

and X is defined by (3.7). The emitted radiation will therefore be of a frequency ν_f (similarly clothed by an electron–hole pair), though $\nu_f \neq \nu$, but $\nu_f < \nu$ if the excitation is generated (Stokes) or $\nu_f > \nu$ if the excitation is already present in the crystal (anti-Stokes), corresponding to a $-$ or $+$ in (4.67). The radiation will be emitted uniformly about the axis of M' with each direction corresponding to a different excitation wave vector K. The value of K can be determined from the conservation of wave vector, and this is discussed in the following section. This process corresponds to an inelastic scattering event at t.

Interaction of radiation with a crystal

The detailed mechanism has been the subject of a number of theoretical studies. Loudon [5, 373] has treated the interaction H_{ER} by perturbation theory and has considered an intermediate process which explains the interaction with phonons. The intermediate process is a virtual process ('that is so in essence or effect although not formally or actually') since it is supposed that the virtual intermediate state arising from H_{ER} involves the excitation of electron–hole pairs, though at the start and at the end of the event the crystal is considered to be in its electronic ground state with all its valence bands full and conduction bands empty. The electron of the electron–hole pair is considered to interact with the phonon (H_{EL}) to explain the Raman process for phonons. Loudon also considered other processes, of secondary importance, involving infrared active phonons. The electromagnetic radiation frequency ν lies in the range (though normally $\nu \neq \nu_0$) of exciton transitions (to be described in chapter 6), and remembering that the polariton range lies between 0 and ν_b polaritons should really be considered as the virtual intermediate state. The electron–hole pair thus becomes represented by the exciton part of the polariton (which now refers to the photon–exciton interaction rather than the photon–phonon interaction) which takes the Coulomb interaction between the electron and hole into account [374, 375] (Loudon assumed free electron–hole pairs). The polariton description was first introduced by Ovander [376–8], extended by Burnstein et al. [379], and Mills and Burnstein [380] who also considered the contribution to the virtual intermediate state of transitions from the valence to the conduction band (Coulomb-correlated free particle–hole pairs), and Bendow [380a]. The principal process is illustrated in fig. 4.8.

4.8 RAMAN SCATTERING GEOMETRIES

Phonons and polaritons (other excitations will be discussed in chapter 6) will be scattered at various angles which can be determined from the conservation of energy and wave vector (fig. 2.23). The scattering geometry has been illustrated in fig. 2.23. Thus

$$K = (k_i^2 + k_f^2 - 2k_i k_f \cos\theta)^{\frac{1}{2}}. \tag{4.68}$$

For phonon or polariton scattering $\nu \approx \nu_f$ and $|k_i| \approx |k_f|$ and therefore (4.68) becomes:
$$K = (\sqrt{2})k_i(1 - \cos\theta)^{\frac{1}{2}}, \tag{4.69}$$

or
$$K = 2k_i \sin\tfrac{1}{2}\theta. \tag{4.70}$$

Raman scattering geometries

In most Raman spectrometers only scattered photons for which $\theta = 90°$ are analysed and for this case

$$K = 2k_1, \qquad (4.71)$$

and, because ν is usually in the visible with k_1 of the order 10^5 cm^{-1}, the region of interest will lie on the right hand side of fig. 4.6.

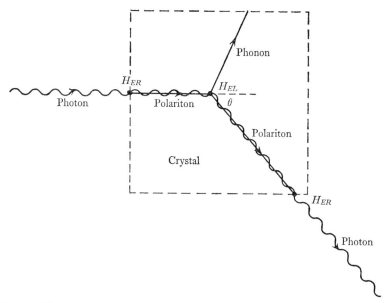

Fig. 4.8. The Raman scattering process showing interaction vertices H_{ER} and H_{EL}.

Thus in the normal Raman arrangement for an infinite crystal.

(i) Longitudinal optical phonons, and polaritons that are almost entirely transverse optical phonons, would be expected to be observed, providing the selection rules of chapter 3 are satisfied

(ii) Since $K > 0$ acoustic phonons (which may be excited as Stokes bands or be present as thermal phonons in the crystal giving rise to anti-Stokes bands) may be observed, though $\nu \mp \nu_t$ is only 1–5 cm^{-1}. Such scattering is known as *Brillouin scattering*.

(iii) Transverse optical phonons that are not infrared active (and will thus have no polariton nature) may be observed, providing the selection rules of chapter 3 are satisfied.

107

Interaction of radiation with a crystal

4.8.1 Selection rules

It is important to remember that the selection rules described in chapter 3 only applied to the case when $K = 0$. The fact that $K > 0$ in the Raman effect sometimes leads to a breakdown in these selection rules. The accuracy of $K = 0$ selection rules obviously depends upon how close the situation at $K = k_1$ approaches that at $K = 0$. Thus it is noted that:

(i) Brillouin scattering is not predicted by $K = 0$ selection rules and occurs only because $K \neq 0$.

(ii) Longitudinal optical phonons (which cannot be infrared active) and infrared inactive transverse optical phonons will have straight line dispersion relations such as those in fig. 4.2 and thus the situation at $K = k_1$ approaches that at $K = 0$ and $K = 0$ selection rules will apply (though the lifting of mode degeneracy by the finite K has been observed [381]).

It is important to note that the problem of incorrect activity predictions from unit cell group analysis discussed in §4.1 for infrared absorption studies does not arise for Raman studies. The unit cell group analysis does apply to the whole crystal as far as α_ξ is concerned. This is because the electronic polarizability expressed by M can be modulated by either the macroscopic transverse electric field of a transverse phonon, or the macroscopic longitudinal electric field of a longitudinal phonon. The unit cell group analysis determines whether such a modulation of M is possible since it will only occur if $|X|$ (3.71) is finite. Thus, for example, for cubic crystals with point groups T, T_h, and T_d the three optical modes for a diatomic crystal belonging to the F representation will be both Raman and infrared active. Thus both longitudinal and transverse phonons would be expected to be observed in Raman studies. Crystals of the zinc blende structure (T_d^2) provide good examples. Thus zinc sulphide gives two bands at 276 and 351 cm^{-1} [382, 283], zinc selenide two bands at 203 and 250 cm^{-1} [384], and zinc telluride two bands at 177 and 208 cm^{-1} [385], and these values are in good agreement with the Lyddane–Sachs–Teller relation (4.13).

4.8.2 Polariton scattering

The region on the right hand side of fig. 4.6, which is important in the normal Raman arrangement, can be seen to exclude polaritons. Polaritons, which have wave vectors on the left hand side of fig. 4.6, can be observed in Raman scattering experiments where $\theta \ll 90°$. For small values of θ

Raman scattering geometries

(known as *near forward scattering*) evaluation of (4.70) shows that K lies in the polariton region.

Since the polariton is a mixed eigenstate of photon and phonon, it thus possesses both a macroscopic electric field [382] and a macroscopic magnetic field [363] and both these fields can be considered under certain conditions to modulate electronic polarizability. Unlike the macroscopic electric field the interaction of the macroscopic magnetic field is not predicted under unit cell group activities. The magnetic field contributes to the electronic polarizability (this contribution has been ignored previously) and a contribution to M can be considered from the varying magnetic moment M_H, arising from the magnetic field H. The interaction of the macroscopic magnetic field of the polariton with the electronic polarizability allows polaritons associated with infrared active modes in centrosymmetric crystals (with therefore no first order Raman spectra) to be observed in near forward scattering experiments [363].

Polariton scattering was first observed in the near forward Raman spectrum of GaP (a diatomic cubic crystal of the zinc-blende structure) by Henry and Hopfield in 1965 [386] (where $\theta = 0$ to $3°$ for the polariton region). Since then a number of uniaxial crystals have been studied in the case of simple diatomic crystals [387] and in the case of more complex crystals with internal vibrations [388–94], the shift of polariton frequency with θ being much more marked than the shift in cubic cases.

Henry and Hopfield [386] suggested that polariton scattering would provide a tunable source of visible (for exciton polaritons) and infrared radiation. This source of radiation will be of low intensity in the normal Raman effect but of high intensity in the stimulated [395] Raman effect. The theory and application of stimulated polariton scattering for the generation of infrared radiation has been discussed [396–8].

4.9 GENERAL WAVES IN VIBRATIONAL SPECTRA

When the lattice dynamics of a three-dimensional lattice was discussed in §2.4 it was seen that, generally, waves in lattices are neither transverse nor longitudinal especially for a very small K. The nature of these general waves and their excitation in vibrational spectra must therefore be considered.

4.9.1 *The infinite cubic crystal*

Although theoretical studies [35] predict that phonons can propagate in directions other than along the main crystallographic directions (§2.4)

Interaction of radiation with a crystal

and therefore be general waves, such general waves that would have a frequency between ν_t and ν_l have not been observed [399]. Electromagnetic radiation propagating in a general direction (Raman studies) excites waves that resolve into transverse and longitudinal modes [399].

Consider, for example, electromagnetic radiation propagating in the xy plane. Radiation propagating along the x and y axes (using the notation

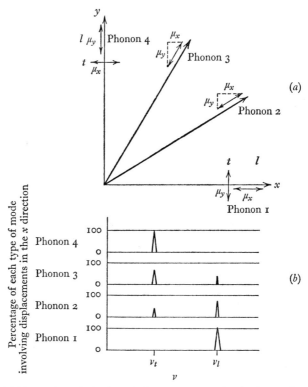

Fig. 4.9. (a) Shows the polarization of phonons propagating in the xy plane. (b) Shows schematically the resolution into longitudinal and transverse phonons that involve displacements in the x direction.

of fig. 4.1) will obviously be able to excite pure longitudinal (though only in Raman studies) and pure transverse phonons. Radiation propagating in general directions in the xy plane excites both longitudinal (only active in the Raman) and transverse phonons as illustrated in fig. 4.9.

This is not to say that no frequencies occur between ν_t and ν_l, but those frequencies that occur in this region are a result of the finite size of the crystal and not the propagation of a general wave.

110

General waves in vibrational spectra

4.9.2 The infinite uniaxial crystal

The situation for uniaxial crystals becomes far more complicated because of the need to consider not only the long range electrostatic forces that lead to longitudinal-transverse splitting but also the short range atomic forces that lead to anisotropy of force constants. These two forces are independent and variations in their relative magnitudes can cause important differences in the nature of the vibrational modes.

The anisotropy of the short range atomic forces causes vibrations involving displacements parallel ($\parallel$) and perpendicular ($\perp$) to the unique axis (which may be taken to be the z axis) to be different. The associated phonons have frequencies denoted $\nu_\parallel$ and $\nu_\perp$ in the notation usually used [5, 400].

TABLE 4.2 *Frequencies of phonons propagating along the main crystallographic axes*

		Phonon polarization direction		
		x	y	z
Phonon propagation direction	x	$\nu_{\perp l}$	$\nu_{\perp t}$	$\nu_{\parallel t}$
	y	$\nu_{\perp t}$	$\nu_{\perp l}$	$\nu_{\parallel t}$
	z	$\nu_{\perp t}$	$\nu_{\perp t}$	$\nu_{\parallel l}$

Modes of the type $\nu_\perp$ are called *ordinary modes* and modes of the type $\nu_\parallel$ are called *extraordinary modes*.

Whatever the relative magnitudes of the two independent forces phonons propagating along the main crystallographic axes will be either of a pure longitudinal or of a pure transverse nature (table 4.2). For electromagnetic radiation propagating in a general direction the nature of the phonons excited will depend upon the relative magnitudes of the two independent forces.

Consider the case when the short range forces are much greater than the longitudinal-transverse splitting. The short range forces will be such that propagation will be forced to occur in the direction of one of the crystal axes. This forced propagation along a crystal axis direction will break down the resolution into transverse and longitudinal phonons illustrated in fig. 4.9 and give a general wave of mixed transverse and longitudinal character. If θ'' is the angle between K and the z axis, then mixed transverse longitudinal phonons in the xz plane will be given by [5]:

$$\nu^2 = \nu_{\parallel t}^2 \sin^2 \theta'' + \nu_{\parallel l}^2 \cos^2 \theta'', \tag{4.72}$$

Interaction of radiation with a crystal

and mixed transverse longitudinal phonons in the xy plane will be given by [14]

$$\nu^2 = \nu_{\perp t}^2 \cos^2 \theta'' + \nu_{\parallel l}^2 \sin^2 \theta''. \tag{4.73}$$

Diagramatically [438] such phonons would be represented by fig. 4.9(a) with all the phonons polarized in the x direction, for mixed transverse longitudinal phonons in the xy plane. If $\nu_{\perp t}$ and $\nu_{\parallel t}$ are infrared active on the basis of unit cell group selection rules mixed modes represented by (4.72) and (4.73) will also be infrared active and, because the mixed modes occur at a frequency between the longitudinal and transverse frequencies, *new modes appear in the spectrum that are not predicted by unit cell group analysis*. The same applies to Raman studies (where, of course, the selection rules apply to longitudinal and transverse modes (see above)).

Consider the case when the short range forces are much smaller than the longitudinal-transverse splitting. Phonons propagating in the xy plane ($\nu_\perp$) will resolve into transverse and longitudinal modes (as in fig. 4.9), but phonons propagating in the xz plane will lead to mixed modes. These mixed modes will be mixtures of a mode of $\nu_\perp$ and $\nu_\parallel$ symmetry which yield a mixed mode of either transverse or longitudinal type. The origin of such mixing is illustrated in fig. 4.10 for transverse phonons. The frequencies of the transverse modes $\nu_{\perp t}$ and $\nu_{\parallel t}$ are different as a result of the anisotropy of the crystal and therefore the mixed modes have a frequency between these two frequencies given by

$$\nu_t^2 = \nu_{\parallel t}^2 \sin^2 \theta'' + \nu_{\perp t}^2 \cos^2 \theta''. \tag{4.74}$$

A diagram similar to fig. 4.10 can be drawn [400] which shows the mixing of the longitudinal modes $\nu_{\perp l}$ and $\nu_{\parallel l}$ and likewise the frequency of the mixed modes will be given by

$$\nu_l^2 = \nu_{\parallel l}^2 \cos^2 \theta'' + \nu_{\perp l}^2 \sin^2 \theta''. \tag{4.75}$$

Thus in the case of these mixed modes *new modes appear in the spectrum that are not predicted by unit cell group analysis* provided that such modes can be active (ν_t might be infrared active and both ν_t and ν_l might be Raman active).

The modes for the general cases where the two independent forces have a variety of relative magnitudes has been described by Loudon [5], but the two limiting cases described above cover many crystals of experimental interest with two atoms in the unit cell.

Merten [401] has extended the theory above to crystals with more than two atoms in the unit cell, and given equations for all the modes for

General waves in vibrational spectra

a general photon propagation direction. The extension of the theory from that of the diatomic unit cell is necessary to allow for the fact that in the multiatomic unit cell there is more than one mode each of the type $\nu_\parallel$ and $\nu_\perp$ which can couple via (4.74) and (4.75). Olechna [402] has analysed the theory for the multiatomic unit cell by a phenomenological treatment

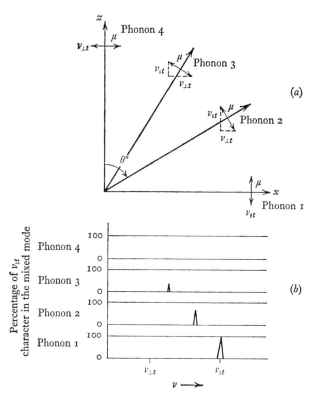

Fig. 4.10. (a) Shows the polarization of phonons propagating in the xz plane. (b) Shows schematically the frequency of the mixed modes of $\nu_{\parallel t}$ character.

and calculated the extraordinary optical phonon frequencies in α-quartz, $CaMoO_4$, $CaWO_4$, Al_2O_3, MgF_2, ZnF_2, trigonal Se, and trigonal Te.

Experimental studies of direction-dependent phonons in uniaxial crystals have included the studies of external modes in wurtzite-type crystals [400, 403], and internal modes in quartz [405, 406, 139, 407, 408], γ-glycine [404], $LiIO_3$ [391, 409], and $CaCO_3$ [409a]. Fig. 4.11 illustrates the mixing of the A and E_1 modes of $LiIO_3$ for phonons propagating in the xz plane (corresponding to (4.74) and (4.75)) where the A mode

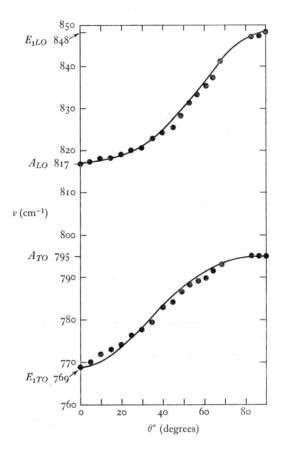

Fig. 4.11. Frequencies of mixed A and E_1 phonons in $LiIO_3$ propagating in the xz plane. The solid line gives the frequencies calculated from (4.74) and (4.75). The diagram is reproduced by permission from Otaguro, Arguello and Porto [409].

corresponds to the ν_1 symmetrical in phase stretching of the IO_3^- ion (there are two molecules in the unit cell), and the E_1 mode corresponds to the ν_3 asymmetrical in phase stretching of the IO_3^- ion.

Loudon has drawn the phonon and polariton dispersion relation for general phonon propagation directions in uniaxial crystals in diagrams analogous to fig. 4.6. The study of polaritons in uniaxial crystals in near forward scattering experiments has led to the observation of a polariton branch crossing a phonon or polariton branch [393, 394]. In fact crossing does not actually [45] occur, the non-crossing rule causing separation of the levels near the point of crossing (this is illustrated in fig. 4.21 for the crossing of a polariton and phonon branch).

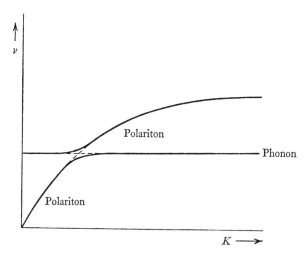

Fig. 4.12. Phonon and polariton dispersion relation near $K = 0$ showing the non-crossing rule for the crossing of polariton and phonon branches with only small interaction between the two branches.

4.9.3 *The infinite biaxial crystal*

There has been less experimental and theoretical work done on biaxial crystals. Merten [410] has described the theory for biaxial (though modifications would have to be made for monoclinic and triclinic crystals) crystals with an arbitrary number of vibrations. Asawa [411] has applied Huang's theory with retardation and has obtained dispersion relations for phonons and polaritons in the case of a diatomic ionic orthorhombic biaxial crystal. In biaxial crystals *all* modes are extraordinary modes, and the uniaxial ordinary mode $\nu_\perp$ becomes replaced by the biaxial extraordinary modes ν_1 and ν_2 ($\nu_\parallel$ becomes ν_3). Thus for biaxial crystals *new modes appear in the spectrum that are not predicted by unit cell group analysis.* The uniaxial case can be shown to be a special case of the biaxial case [401]. Experimental studies have detected longitudinal-transverse splitting in such systems [305, 381, 412].

Interaction of radiation with a crystal

4.10 THE EFFECTS OF FINITE SAMPLE SIZE

So far the cases of infinite (or very large) samples have been considered. The effect of finite crystal size on the lattice dynamics has already been discussed (§2.7) where it was seen that the modes developed a size and shape dependence. The effect that this size and shape dependence has upon the optical properties is now discussed. For a cubic crystal three types of mode were found: transverse modes, longitudinal modes, and surface modes. The origin of surface modes can be understood by returning to Huang's theory discussed in §4.3. In this theory u was split into solenoidal and irrotational parts ((4.5) and (4.6)). Barron [413] has shown that this step involves the assumption that the crystal is infinite, or obeys cyclic boundary conditions, and Rosenstock [35] has shown that the use of cyclic boundary conditions is only valid for crystals with no long range forces. Thus in finite crystals with such long range forces (which will have a Coulomb nature) the theory must be modified. Barron [413] showed that u in such cases should be split into solenoidal, irrotational and a part that is both solenoidal and irrotational:

$$u = u_t + u_l + u_s, \tag{4.76}$$

where

$$\left.\begin{array}{r}\nabla \cdot u_s = 0, \\ \nabla \times u_s = 0.\end{array}\right\} \tag{4.77}$$

The modes given by u_s are known as *surface modes*.

The degree to which the surface modes extend into the bulk, and thus cause size and shape dependence, depends upon the long range Coulomb forces. Long range Coulomb forces will, of course, be much weaker in covalent crystals, and they will be much less important for internal vibrations (where intramolecular forces are much greater than intermolecular forces) than for external vibrations.

The effect that these surface modes have on the vibrational spectra of finite crystals has been illustrated by evaluating their frequencies for ionic crystals of various geometries.

4.10.1 *Crystals of extremely small size*

It is easier to first examine the case of a sample of extremely small size. If the sample size is much smaller than the wavelength of the electromagnetic radiation that excites the vibrations, then the photon part of the polariton can be ignored and pure phonons considered. This neglect of retardation makes the theory *much simpler*

The effects of finite sample size

(i) *The cubic ionic crystal slab*

Consider a slab of the geometry shown in fig. 4.13. The frequencies of the transverse, longitudinal, and surface modes for the slab may be calculated most simply by the method of Ruppin and Englman [414], or by considering the slab as a limiting case of a concentric hollow cylinder [415], or by the complicated method of Fuchs and Kliewer [416]. In the very thin slab limit considered here the electric field of the electromagnetic radiation does not couple with the bulk modes which are identical

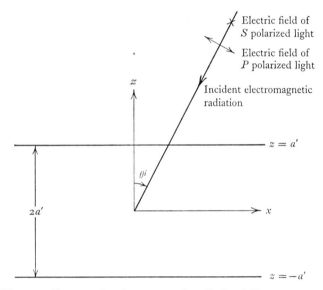

Fig. 4.13. Geometry for electromagnetic radiation falling on a crystal slab of thickness $2a'$.

to the transverse and longitudinal optical modes of the infinite crystal, but only couples with the surface modes. The variation of the surface mode frequencies with the thickness of the slab, calculated by the methods described above, is given for a crystal in air in fig. 4.14. It can be seen from fig. 4.14 that the surface mode frequency, which lies between ν_l and ν_t, splits into a high and a low frequency mode which approach these two frequencies in the thin slab limit. The high frequency surface mode can be shown [416] to have atomic displacements in the z direction (i.e. of longitudinal character), and the low frequency surface mode to have atomic displacements in the x or y direction (i.e. of transverse character),

Interaction of radiation with a crystal

both modes being characterized by the wave vector K_x. In the thin slab limit the excitation of these modes for incident radiation falling on the slab as shown in fig. 4.13 (which may be P (polarized in the xz plane) or S (polarized in the xy plane)) will depend upon the angle θ^i. If $\theta^i = 0$ only the transverse character surface modes can be excited. If $\theta^i \neq 0$ there will be a component of the P polarized in the z direction and the longitudinal character surface mode will also be able to be excited. Since this

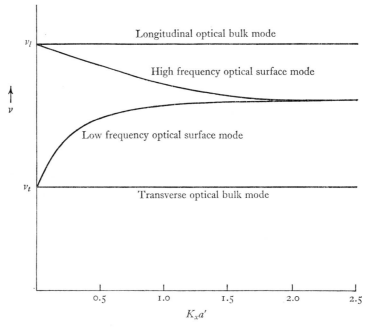

Fig. 4.14. Diagram illustrating the type of dependence that the optical surface mode frequencies have on slab thickness ($2a'$). Bulk modes have no meaning for thin slabs since the surface modes penetrate a number of atomic layers into the bulk.

mode is a *localized mode* the activity considerations illustrated in fig. 4.1 for longitudinal vibrations of an infinite crystal no longer apply and such a mode can be excited. It follows from the above that absorption of P polarized light leading to longitudinal character surface modes will be proportional to $\sin^2 \theta^i$.

Berreman [417] has considered the absorption by thin flat films of cubic ionic crystals, and concluded that in both transmission and reflection studies bands characteristic of longitudinal optic modes of infinite crystals can be excited by non-normal P polarized radiation. These bands

The effects of finite sample size

are thus the high and low frequency modes of the thin slab limit discussed above. Fig. 4.15 illustrates the results of Berreman's studies on thin films of lithium fluoride with the observation of the longitudinal character surface mode at 675 cm^{-1} for P polarized light at non-normal incidence (compare with the calculated value (table 4.1)). The results of some other experimental studies leading to the longitudinal optical frequency in the thin slab limit are listed in table 4.3.

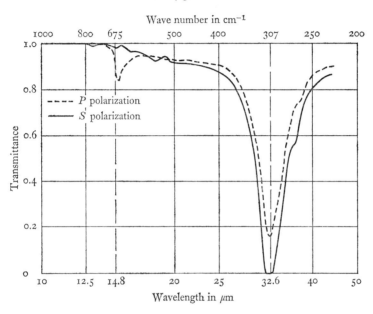

Fig. 4.15. Observed transmittance at room temperature of S polarized and P polarized radiation by a LiF film 0.20 μ thick deposited at 265 °C on collodion, relative to that by uncoated collodion. Radiation incident in a cone from 26 to 34 degrees. Reproduced by permission from D. W. Berreman [417].

TABLE 4.3 *Longitudinal optical frequencies obtained in non-normal incidence P polarized radiation experiments with films in the thin slab limit*

Crystal	Space group	ν_1 (cm^{-1})	Reference
CaF$_2$	O_h^5	456	419
SrF$_2$	O_h^5	366	419
BaF$_2$	O_h^5	319	419
AgCl	O_h^5	196	412
AgBr	O_h^5	138	412
ZnSe	T_d^2	255	420
ZnTe	T_d^2	206	420
CdS	C_{6v}^4	302	421

Interaction of radiation with a crystal

(ii) *The cubic ionic crystal sphere*

This case was first considered by Fröhlich [418] for a diatomic crystal with a radius large compared with the lattice constant, but small compared with the wavelength of the electromagnetic radiation. Fröhlich found that a vibrational mode with a polarization homogeneous throughout the spherical volume could exist, and this mode had a frequency ν_F intermediate between ν_t and ν_l, given by

$$\frac{\nu_F^2}{\nu_t^2} = \frac{\epsilon_0 + 2n_m}{\epsilon_\infty + 2n_m}, \qquad (4.78)$$

where the sphere is in a medium of refractive index n_m. The surface modes for the sphere may be calculated [414] and will be given by

$$\frac{\nu_{s1}^2}{\nu_t^2} = \frac{\epsilon_0 + n_m(l+1)/l}{\epsilon_\infty + n_m(l+1)/l} \quad (l = 1, 2, 3, ...), \qquad (4.79)$$

where clearly the lowest surface mode ($l = 1$) is the Fröhlich mode ($\nu_{s_1} = \nu_F$). The lowest surface mode has a constant amplitude over the whole volume of the sphere, and the higher modes an amplitude which decreases as r^{l-1} with increasing distance from the surface of the sphere. Fig. 4.16 illustrates the dependence of ν_s on n_m for NaCl.

(iii) *The cubic ionic crystal cylinder*

Expressions for the surface mode frequencies have been given by Ruppin and Englman [414, 415].

4.10.2 *Crystals of small to large size*

For all finite sized crystals, except those of extremely small size, the photon part of the polariton must be considered.

When a crystal of finite size is considered it is possible for the polariton to be totally internally reflected and thus become trapped in the crystal. This leads to two types of polariton modes.

(i) *Radiative modes*, these are polaritons that are identical to those discussed previously, and are thus virtual modes with oscillating fields outside the crystal (where all the phonon energy is transferred to photon energy, and the photon has the same energy as the incident photon).

(ii) *Non-radiative modes*, these are polaritons which, for certain crystal shapes, are totally internally reflected and thus trapped in the crystal. The non-radiative modes are thus true normal modes in the sense that,

The effects of finite sample size

in the absence of damping, they persist forever after being initially excited. The non-radiative modes thus have no direct interaction with incident electromagnetic radiation.

The possible existence of non-radiative modes will clearly depend upon the crystal having the necessary shape for total internal reflection to occur.

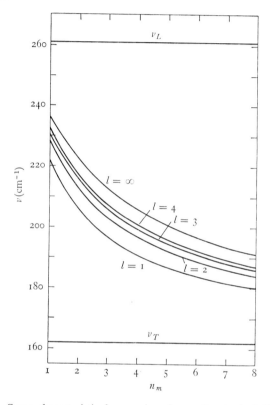

Fig. 4.16. Some characteristic frequencies of a small spherical diatomic crystal as a function of the dielectric constant n_m of the medium. The (longitudinal and transverse) bulk frequencies and the surface phonon frequencies are shown. The parameters in the figure are appropriate to NaCl. Reproduced by permission of Ruppin and Englman [414].

(a) Crystal slab. Radiative modes [422] occur when $K > k_i^m$ and non-radiative [423] modes when $K < k_i^m$, where k_i^m is the photon wave vector in the air in which the slab is placed (*not* the photon wave vector for the lattice medium illustrated in fig. 4.2). The situation is illustrated in fig. 4.17. When the slab is placed with one face in one medium and the

Interaction of radiation with a crystal

other face in another medium some modes can radiate on one side of the slab and some on both sides [414].

(b) Crystal sphere. All the modes are radiative.

(c) Crystal cylinder. The polaritons separate into radiative modes and non-radiative modes in the same way as for a crystal slab in air.

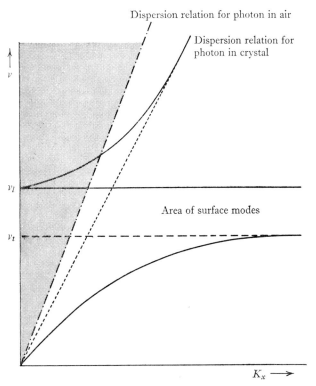

Fig. 4.17. Dispersion relation for photons, phonons, and polaritons, showing the area of radiative modes (shaded) in the case of a crystal slab or crystal cylinder (when K_x is taken parallel to the cylinder axis) in air.

To predict the spectroscopic properties it is therefore necessary to obtain the frequencies of the radiative polariton modes. It was seen in §4.8 how both the macroscopic electric field and the macroscopic magnetic field of the polaritons need to be considered. These fields may be obtained by an extension of Huang's theory to the case of a finite crystal. To obtain the dispersion relations (3.63) is operated on with the curl operator:

$$\nabla \times \nabla \times \boldsymbol{E} = -\frac{2\pi\nu\mathrm{i}}{c}[\nabla \times \boldsymbol{H}], \qquad (4.80)$$

The effects of finite sample size

where H is given by

$$H = H_0 \cdot e^{-2\pi i \nu t}. \tag{4.81}$$

Substituting for $\nabla \times H$ from (3.62), with E and P given by (4.8) and (4.10), and then substituting for P from (4.12), (4.80) becomes

$$\nabla \times \nabla \times E = \frac{4\pi^2 \nu^2}{c^2} \left[1 + 4b_{22} + \frac{4b_{12} b_{21}}{-b_{11} - 4\pi^2 \nu^2} \right] E, \tag{4.82}$$

and since

$$\nabla \times \nabla \times E = \nabla(\nabla \cdot E) - \nabla^2 E, \tag{4.83}$$

and for transverse modes

$$\nabla \cdot E = 0, \tag{4.84}$$

then (4.82) becomes (introducing the dielectric constant via (4.13)):

$$\left[\nabla^2 + \frac{4\pi^2 \nu^2}{c^2} \right] E = 0. \tag{4.85}$$

Similarly for the magnetic field:

$$\left[\nabla^2 + \frac{4\pi^2 \nu^2}{c^2} \right] H = 0, \tag{4.86}$$

where in the special case of an infinite crystal or one subject to cyclic boundary conditions (4.85) and (4.86) become:

$$\epsilon_\nu = \frac{K^2 c^2}{4\pi^2 \nu^2}. \tag{4.87}$$

For finite crystals (4.85) and (4.86) must be solved for E and H inside and outside the crystal. These equations may be solved for rectangular, spherical, circular-cylindrical elliptic-cylindrical, parabolic-cylindrical and conical coordinate systems corresponding to the various crystal shapes. For slabs, spheres and circular cylinders the transverse solutions of (4.85) and (4.86) are [414]

$$M = \nabla \times a_1 w \psi, \tag{4.88}$$

$$N = \frac{1}{K} [\nabla \times \nabla \times a_1 w \psi], \tag{4.89}$$

where ψ is the solution of the scalar equation:

$$(\nabla^2 + K^2) \psi = 0 \tag{4.90}$$

and $a_1 = a_z$, and $w = 1$ for slabs or circular cylinders, and $a_1 = a_r$, and $w = r$ for spheres.

Interaction of radiation with a crystal

The virtual modes can therefore be expressed in terms of the electric and magnetic fields associated with them. The virtual modes have a real wave vector and a complex frequency of the form:

$$\nu = \nu'_r + i\nu'', \qquad (4.91)$$

where ν'_r is the real frequency. To obtain these frequencies ψ is obtained for the particular crystal shape [414], (4.88) and (4.89) solved to yield the electric and magnetic fields, and the appropriate boundary condition applied. The virtual modes are classified.

(i) E is either of the form of (4.88) and H of the form of (4.89) or vice versa. This leads to the description *electric mode*, given the symbol E, or *magnetic mode*, given the symbol M. Thus *electric modes* are characterized by $E \propto N$ and $H \propto M$. For a sphere the magnetic field associated with these modes has no radial component. For a cylinder the magnetic field associated with these modes has no component along the cylinder axis. For a slab the magnetic field associated with these modes has no component in the xz plane and the modes are known as *P polarized modes* (since E lies in the xz plane).

Magnetic modes are characterized by $E \propto M$ and $H \propto N$. The electric field lacks the same components in the various crystal shapes as the magnetic field lacked in the electric modes above. For a slab the modes are known as *S polarized modes* (since E lies in the xy plane).

(ii) The surface modes, given the symbol S, lie in the range $\nu_t < \nu_s < \nu_l$. The bulk modes are classified in two groups.

(a) Low frequency modes, given the symbol L, form a series of frequencies below ν_t converging to ν_t as the sample dimensions become smaller.

(b) High frequency modes, given the symbol H, form a series of frequencies above ν_l which do not converge and spread out to $\nu = \infty$.

The details of the calculation of the virtual mode frequencies can be found elsewhere [414, 422–7] and only the results will be given here.

(i) *The crystal slab*

In this case there are *no* radiative surface modes, and thus only the radiative bulk modes are considered. Since the slab possess translational symmetry in the x direction there will be an $e^{iK_x x}$ x dependence, and ψ will be given by

$$\psi = e^{iK_x x} e^{i\beta' z}, \qquad (4.92)$$

The effects of finite sample size

inside the slab and

$$\psi = e^{iK_x x} e^{i\beta'_0 z}, \qquad (4.93)$$

outside the slab (in air), where

$$\beta' = \left(n\frac{4\pi^2 \nu^2}{c^2} - K_x^2\right)^{\frac{1}{2}}, \qquad (4.94)$$

and β'_0 is given by (4.94) with n replaced by n_m. The frequencies can be shown to be given by

$$\tan \beta' a' = -in\frac{\beta'_0}{\beta'}, \qquad (4.95)$$

$$\cot \beta' a' = in\frac{\beta'_0}{\beta'}, \qquad (4.96)$$

where (4.95) and (4.96) apply to the P polarized modes of a crystal slab in air. The S polarized modes are also given by (4.95) and (4.96) with ϵ replaced by 1. The P polarized case is that usually studied, though it has been shown [423] that a reasonable idea of the structure of the S polarized modes can be obtained from the properties of the P polarized modes. The differences in optical properties for the two polarizations arise from the replacement of n by 1 for the S polarized case leading to differences in maxima and minima of reflection [427].

The virtual modes of a slab are characterized, in addition to the features above, by C and T depending upon whether they are solutions of (4.96) or (4.95), and by integers, l, given by the limit of (4.95) and (4.96), where

$$\beta' a' = \frac{l\pi}{2}. \qquad (4.97)$$

The number of virtual bulk modes can be seen to be infinite, since there will be an infinite number of high frequency modes. The determination of the optical properties depends upon determining the important modes that lie in the region of ν_t.

Let the thickness of the slab be denoted by W where

$$W = \frac{4a'\pi\nu_t}{c}, \qquad (4.98)$$

and the frequency of the mode be given in terms of Ω, where

$$\Omega = \nu/\nu_t. \qquad (4.99)$$

Fig. 4.18 shows the calculated absorption and fig. 4.19 the calculated reflection for a relatively thick slab with $W = 10$. The convergent series of low frequency modes can be seen converging to ν_t with l corresponding

Interaction of radiation with a crystal

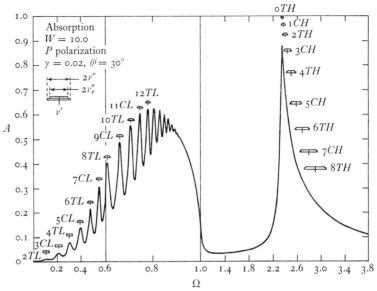

Fig. 4.18. Absorption by a slab with $W = 10$. Reproduced by permission from Fuchs, Kliewer and Pardee [427].

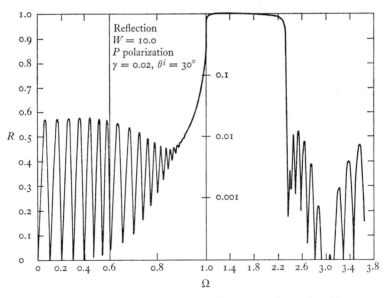

Fig. 4.19. Reflection by a slab with $W = 10$. Reproduced by permission from Fuchs, Kliewer and Pardee [427].

The effects of finite sample size

alternatively to C and T modes, and the infinite number of high frequency modes can be seen spreading out (with gradually decreasing importance in terms of absorption and reflection) to $\Omega = \infty$. When the slab is made thinner (fig. 4.20 shows the case for $W = 1$), the low frequency modes converge more rapidly, and the high frequency modes spread out more rapidly. Finally in the thin slab limit (fig. 4.21 shows the case for absorption and fig. 4.21 the case for reflection and transmission for $W = 0.1$),

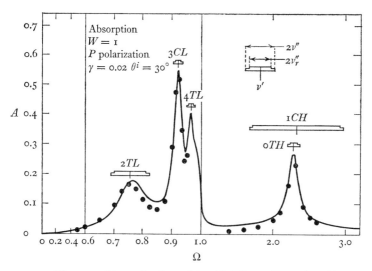

Fig. 4.20. Absorption by a slab with $W = 1$. Reproduced by permission from Fuchs, Kliewer and Pardee [427].

two modes occur. One corresponds to the low frequency limit at ν_t, and the other to the high frequency mode oTH at ν_l, or rather very nearly at ν_l where

$$\left(\frac{\nu_{0TH}}{\nu_t}\right)^2 = \frac{\epsilon_0 \epsilon_\infty - \epsilon_0 W^2 \tan^2 \theta^i}{\epsilon_\infty^2 - \epsilon_0 W^2 \tan^2 \theta^i}, \qquad (4.100)$$

which, using (4.81) becomes nearly ν_l when W is very small. The results for the thin slab thus agree with those of the previous section when only the phonon part of the polariton was considered. It is necessary to use the full polariton treatment if the widths of the modes are to be calculated [421]. The way in which the widths may be calculated (the calculated widths are illustrated in figs. 4.18–4.21) will be left to the next chapter.

Interaction of radiation with a crystal

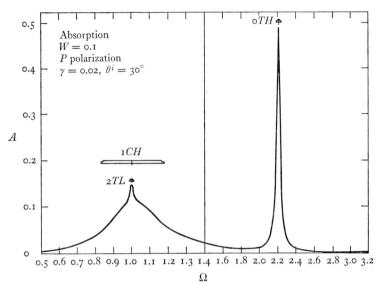

Fig. 4.21. Absorption by a slab with $W = 0.1$. Reproduced by permission from Fuchs, Kliewer and Pardee [427].

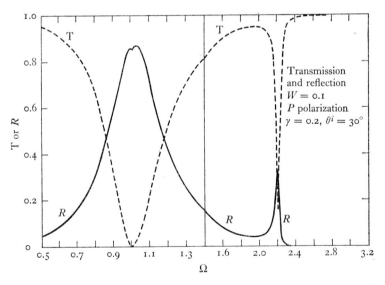

Fig. 4.22. Transmission and reflection by a slab with $W = 0.1$. Reproduced by permission from Fuchs, Kliewer and Pardee [427].

The effects of finite sample size

(ii) *The crystal sphere*

The case of a sphere is more complicated since *all* the modes are radiative. All the surface modes are thus radiative and additional modes will appear between ν_t and ν_l. It has been seen (4.79) that there is one surface mode for each l value but it can be shown that there is an infinite number of bulk modes. For a sphere of general finite size, the polariton surface modes are transverse, though when the radius approaches zero they are

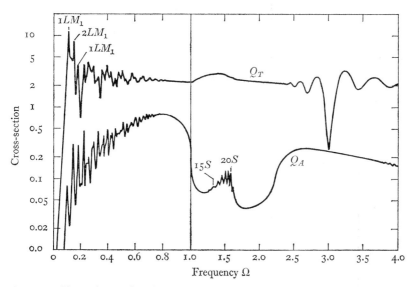

Fig. 4.23. Absorption and extinction cross-sections (Q_A and Q_T) of a sphere with $W = 10$. Reproduced by permission from Fuchs and Kliewer [424].

given by the phonon surface mode condition (4.77). It has been noted (4.79) that when $l \to \infty$ the field of the surface mode becomes increasingly localized at the surface of the sphere, and for such a case, the surface mode has a constant frequency of

$$\frac{\nu_{s\infty}^2}{\nu_t^2} = \frac{\epsilon_0 + n_m}{\epsilon_\infty + n_m}, \qquad (4.101)$$

thus the surface modes form a series which converges to a limiting frequency $\nu_{s\infty}$.

Figs. 4.23 to 4.25 show the calculated absorption and extinction cross-sections for various sphere sizes $W = 10$, $W = 1$, and $W = 0.1$ (where the sphere radius replaces a' in (4.98)). The high and low frequency modes behave in a similar way to those of a slab when $W \to 0$ (the subscript

Interaction of radiation with a crystal

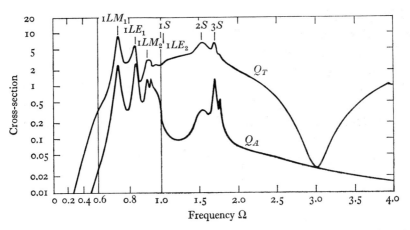

Fig. 4.24. Absorption and extinction cross-sections (Q_A and Q_T) of a sphere with $W = 1$. Reproduced by permission from Fuchs and Kliewer [424].

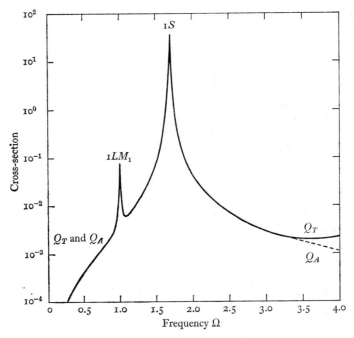

Fig. 4.25. Absorption and extinction cross-sections (Q_A and Q_T) of a sphere with $W = 0.1$. Reproduced by permission from Fuchs and Kliewer [424].

The effects of finite sample size

labels the infinite number of modes of the same l for a given electric or magnetic mode) the low frequency modes converging to ν_t and the high frequency modes now diverging to an infinitely high frequency. Unlike the slab case, however, the surface modes appear between the high and low frequency modes with frequencies converging to $\nu_{s\infty}$. In the case of a very small ($W = 0.1$) sphere it can be seen that the spectrum consists of a $1LM_1$ mode at ν_t and the $1S$ surface mode which is the Fröhlich mode at ν_F. In the limit of an extremely small sphere only the $1S$ mode appears as predicted in the previous section when only the phonon part of the polariton was considered.

Ruppin and Englman [414] illustrate the effect of the surrounding medium on the modes in fig. 4.25 for media of varying dielectric constant.

(iii) *The crystal cylinder*

The case of the cylinder is intermediate between that of a slab and a sphere. Like the slab, radiative and non-radiative modes occur, but unlike the slab radiative surface modes are possible, though only of the magnetic type. In the extremely small cylinder limit [414] the $1S$ surface mode can be excited for perpendicular polarization ($\boldsymbol{E} \perp \boldsymbol{x}$) and the low frequency mode series limit frequency (ν_t) for parallel polarization ($\boldsymbol{E} \| \boldsymbol{x}$).

The spectra of multiatomic crystals are similar to those of diatomic crystals discussed above, with the structure repeating itself for each frequency ν_t. Fig. 4.26 illustrates the case of $KMgF_3$ calculated by Ruppin and Englman [414].

4.10.3 *Shape effects for internal vibrations of covalent cubic crystals of extremely small size*

Fox and Hexter [428] have considered the shape dependence of internal vibrations in covalent crystals for extremely small cubic crystals where the particle size is large compared with the lattice constant, but small compared with the wavelength of the electromagnetic radiation. The interaction between one molecule and the molecules of a single sublattice by means of transition dipole–transition dipole interactions is expressed by a Hamiltonian $\mathcal{H}$ given by

$$\mathcal{H} = \mathcal{H}_R + \mathcal{H}_S, \quad (4.102)$$

where $\mathcal{H}_R$ is the Hamiltonian that leads to correlation field splitting,

Interaction of radiation with a crystal

and $\mathscr{H}_S$ is the Hamiltonian that leads to shape splitting. $\mathscr{H}_S$ does not affect inactive states and only contains interactions of range r^{-3} and greater. The symmetry of $\mathscr{H}_R$ is given by the unit cell group and the symmetry of $\mathscr{H}_S$ by the group of operations which take the sample surface into itself. The consideration of $\mathscr{H}$ rather than $\mathscr{H}_R$ leads to the following possible effects.

(i) 'Shape splittings' may occur because degeneracies under $\mathscr{H}_R$ symmetry may be removed for $\mathscr{H}$ symmetry, and the introduction of $\mathscr{H}_S$ may mix states which are separated for $\mathscr{H}_S$.

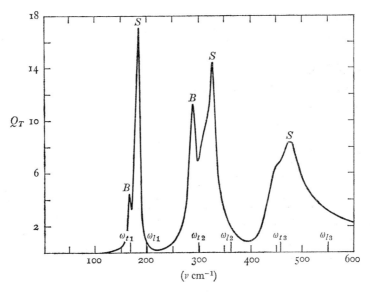

Fig. 4.26. Extinction cross-section (Q_T) of spherical particles (of radius 4 μm) dispersed in a polythethylene medium. Bulk (B) and surface (S) modes are shown. Reproduced by permission from Ruppin and Englman [414].

(ii) The dipole lattice may have a higher symmetry than that of the crystal itself (for example in a crystal of unit cell group C_1 the Hamiltonian has a symmetry C_i, since all interactions are quadratic in the dipole moments).

For triply degenerate F states, splitting occurs into a non-degenerate (polarization parallel to the unique axis of the sample ellipsoid) and a doubly degenerate state (polarization perpendicular to the unique axis of the sample ellipsoid) the splitting, ΔE being given by

$$\Delta E = \tfrac{3}{2}ga''^{-3}P_F^2, \qquad (4.103)$$

where g is a parameter which depends on the axial ratio of the ellipsoid and has the values:

$$g = 0 \text{ for a sphere,} \tag{4.104}$$

$$g = -\frac{8\pi}{3} \text{ for a thin slab,} \tag{4.105}$$

$$g = \frac{4\pi}{3} \text{ for a long needle.} \tag{4.106}$$

a'' is the length of the unit cell side, and P_F is the magnitude of the transition dipole moment to an F state where

$$P_F^2 = \frac{1}{6}\left(\frac{\partial \mu_\xi}{\partial Q_k}\right)^2 \frac{h}{\pi^2 \nu}. \tag{4.107}$$

Using the theory above Fox and Hexter made comparisons with experiment. The importance of interactions other than transition dipole–transition dipole interactions (e.g. multipole interactions, repulsive interactions) has been mentioned.

A linear response theory has been applied to the propagation of photons through molecular crystals [429], and the dipole response of a cubic molecular crystal found to have two parts, one (the virtual response) that is not excited by light [430] and the other (the real response) that is excited by light and is surface dependent. It has been shown [431] that the propagation of photons by this theory agrees with the polariton (the exciton–photon type) theory (described in §4.7) only if Umklapp processes (§5.33) are included.

The consideration of the covalent crystal case, where long range Coulomb forces are very weak, is thus complimentary to the size effects in predominantly ionic crystals.

4.10.4 *Uniaxial crystals*
Ruppin and Englman [414] have explained how the equations can be modified to cope with uniaxial crystals.

4.10.5 *Experimental infrared transmission studies*
Many experimental studies involve samples in the form of powders often suspended in mulls of liquids or in discs of solids (e.g. polyethylene discs for far infrared studies). These powder samples can be analyzed using the theory applicable to a spherical crystal, though further complications due to scattering and interference occur (to be discussed in the next chapter). *Many experimental studies on powders note a shift to frequencies*

Interaction of radiation with a crystal

higher than v_t and a general spectral sharpening as the powders are more finely ground, due to approach to the limit of an extremely small sphere with one $1S$ mode (and the frequency of this mode depends upon the dielectric constant of the medium in which the powder is suspended).

Ruppin and Englman [414] list a number of examples of surface modes determined experimentally for very small diatomic cubic crystals. For larger samples the experimental observation of the theoretically predicted fine structure depends upon the uniformity of size and shape of the

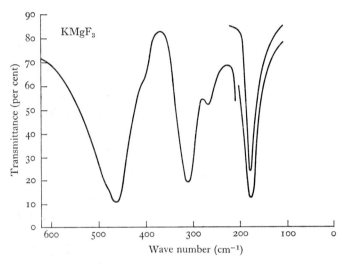

Fig. 4.27. Infrared spectrum of powered KMgF$_3$ in polyethylene. Reproduced by permission from Hunt, Perry and Ferguson [434].

samples, spectrometer resolution and energy, and the lack of complications due to scattering and interference. Thus Fuchs and Kliewer [424] obtained only partial agreement with the experimental results [432] for UO_2 particles in polythene. Martin [433] obtained better agreement with experimental studies on KCl crystals and his results show bulk and surface modes where the ratio of bulk to surface mode absorption is seen to increase with crystal size. The transmission spectrum of powdered KMgF$_3$ dispersed in polyethylene films [434], shown in fig. 4.27, gives good agreement with the calculated spectrum (fig. 4.26). Examples of other multiatomic cubic crystals are given by Ruppin and Englman [414]. Ruppin and Englman [414] also list a number of examples of the observation of the $1S$ surface mode (giving good agreement with the mode calculated from (4.78)) in finely ground powders of uniaxial crystals.

The effects of finite sample size

4.10.6 *Experimental Raman studies*

The size and shape effects discussed above would be expected to be observable in Raman studies. In these studies the retardation effects caused by phonon–photon interactions discussed above will not be observed and the results calculated for the phonon part of the polariton can be applied.

Surface modes can be detected for the slab geometry (fig. 4.13) where right angle backwards (where the scattered beam is detected at 90° to the incident beam in the $+z$ direction (fig. 4.13)) scattering has proved most convenient. Ruppin and Englman [414] have described the scattering geometry for surface modes where it can be seen from fig. 4.13 that

$$K_x = k_i \sin\theta^i - k_f \cos\theta^i. \qquad (4.108)$$

As θ^i increases surface phonons with smaller K_x values are studied, and Rupin and Englman [414] illustrate the case for infrared and Raman active diatomic cubic crystals of the zinc-blende structure, when fig. 4.14 applies, and four Raman lines should be seen. Acoustic surface phonons have also been observed by a similar method [435, 436].

Optical surface phonons would be expected to occur in those studies of powders (where complications occur due to the elastic scattering of the exciting line by the powder particles) and small crystalline slabs where the bulk phonons studied were both infrared and Raman active, and the theory above used to analyze the results.

4.10.7 *Experimental electron scattering studies*

Optical surface phonons have been detected by electron scattering studies using high and low energy electrons. Charged particles would be expected to excite the longitudinal optical phonon, but optical surface phonons can also be excited. Boersch, Geiger and Stickel [437, 438] using 25 keV electrons, giving a resolution of 323 cm^{-1}, detected a strong energy loss spectrum (408 cm^{-1}) due to the low frequency optical surface mode (4.14), and a weak energy loss spectrum for the high frequency optical surface mode and the longitudinal optical surface mode, for thin films of ionic crystals. Ibach [439] using slow electrons of only 7.5 eV, giving a resolution of less than 161 cm^{-1}, obtained an energy loss spectrum from uniaxial zinc oxide crystals due to optical surface phonons at 556 cm^{-1} and 543 cm^{-1}. The theory has been discussed [439a]

5 Second order vibrational spectroscopic features

5.1 INTRODUCTION

Having fully examined the first order vibrational spectroscopic features, the second order features can be examined. Those second order features that allow the actual appearance of the spectrum to be understood (the band shapes and intensities) are discussed together with second order features such as the effects of temperature, pressure, applied fields, and defects.

5.2 THE WIDTH AND TEMPERATURE DEPENDENCE OF FUNDAMENTALS

5.2.1 *Width arising from the damping factor γ*

In §4.6 it was seen how the presence of anharmonic terms and/or non-linear dielectric nature, together represented by γ, caused phonons to couple and no longer be mutually independent [440]. Anharmonic terms cause the potential energy expression to be modified by nth order terms, thus the potential energy for a diatomic lattice expressed by (3.28) becomes modified:

$$\text{potential energy} = \tfrac{1}{2}\sum_{k} \lambda_k Q_k^2 + \sum_{kk'k''} \lambda_{kk'k''} Q_k Q_{k'} Q_{k''} + \cdots, \quad (5.1)$$

where k, k', and k'' refer to different normal coordinates, hence coupling different phonons. A non-linear dielectric nature causes nth order non-linear terms in the crystal dipole moment, thus the dipole moment of a diatomic lattice per unit volume expressed by (4.37) becomes modified:

dipole moment per unit volume in the direction ξ

$$= A\left[\left|\frac{E_l}{E}\right|\frac{1}{\mu'}\right]^{\frac{1}{2}} [q^* Q_k + \sum_{k'k''} q'_{k'k''} Q_{k'} Q_{k''} + \sum_{k'k''k'''} q''_{k'k''k'''} Q_{k'} Q_{k''} Q_{k'''} + \cdots],$$

$$(5.2)$$

where Q_k represents the $K = 0$ normal coordinate in the ξ direction, and $q'_{k'k''}$, $q''_{k'k''k'''}$ etc. represent the higher order contributions of the distortion moments, with k, k', k'', and k''' referring to different normal coordinates,

Width and temperature dependence

hence coupling different phonons. The $(n+1)$th order term in the potential energy and/or the nth order term in the dipole moment makes possible n-phonon transitions.

The damping factor γ causes the observed spectral line to have a finite width. If w is the half-width of the spectral line arising from γ, then [441]:

$$w = \frac{\gamma}{2\pi}. \qquad (5.3)$$

The value of γ (or the frequency dependent γ'/ν_0 in the quantum mechanical treatments (§4.6)) has been obtained theoretically in order to calculate a value of w. Cowley [372] has used a many-body theory with quantum mechanical treatment, Maradudin and Wallis [368] use a quantum statistical approach, Kleinman [442] obtains γ for the case where the principal relaxation mechanism is the anharmonic perturbation, Mitskevich [369] and Wehner [371] use quantum mechanical methods, and these theoretical studies and various experimental studies have been reviewed [12, 443]. Plendl [444] has used a completely different approach analyzing the damping due to collisions between the atoms during vibrations by kinetic theory comparing the w (5.3) so obtained with the experimental results for some eighty, largely diatomic, crystals.

5.2.2 Width arising from virtual modes

In §4.5 and §4.6 it was seen how the polaritons (the radiative type) were the virtual modes responsible for the emission of radiation from the crystal, and how absorption required phonon–phonon interaction leading to a reduction of polariton energy flux. This phonon–phonon interaction discussed above was seen to lead to a finite lifetime of the phonon part of the polariton and thus a spectral line width w. It must be remembered that the polaritons of interest (the radiative modes) are virtual modes and thus have a finite lifetime, and the contribution of the resulting radiative width (inverse of lifetime) caused by the flow of energy out of the sample to the observed spectral line width must be considered. The total width will be given by

$$\text{total width} = |2\nu''|, \qquad (5.4)$$

where ν'' is given by (4.91), and ν is considered to apply to the case where $\gamma = $ finite. If ν with $\gamma = 0$ is represented by ν_r and defining:

$$\Delta\Omega'' = \nu'' - \nu_r'', \qquad (5.5)$$

then

$$\text{total width} = |2\nu''| = |2\nu_r''| + |2\Delta\Omega''|, \qquad (5.6)$$

Second order vibrational features

where $|2\nu_r''|$ represents the radiative width, and

$$|2\Delta\Omega''| = w. \qquad (5.7)$$

In the limit when the sample size is zero the radiative width is zero, but it has been shown [427] for a finite crystal slab that one requires a very thin crystal in order that the radiative contribution to the width is negligible compared to w and therefore one must be cautious before calculating γ from the measured width of an absorption peak. w and the radiative width are shown in figs. 4.18, 4.20, and 4.21. Priox and Balkanski [421] have compared the width calculated from (5.6) with experimental values obtained for thin films.

5.2.3 *Temperature dependence of the width and frequency of fundamentals*

The width and observed frequency ν' of a fundamental spectral line will depend upon temperature. γ (and thus w) is temperature dependent and the treatments for the calculation of γ discussed above and other studies [445–51] allow this temperature dependence to be calculated, γ being found to increase with increasing temperature. The actual temperature dependence of γ depends upon the same and may vary as T, $T^{\frac{3}{2}}$ [12, 443], or T^2 [452] and may be explained by considering various cubic and quartic anharmonic terms in the crystal potential energy [450, 451, 453].

The temperature dependence of the frequency ν' of the fundamental is largely due to pseudo-harmonicity. Pseudo-harmonicity (or quasi-harmonicity) arises because anharmonic forces cause the volume of a crystal to become temperature dependent [442, 454]. The change in the separation of the atoms, that results from a temperature dependent volume, is analysed by harmonic forces and this change gives rise to a temperature shift. This shift will move ν' to higher frequencies on cooling.

Pseudo-harmonicity assumes that ν' is independent of temperature at constant volume, which is not correct. The presence of anharmonic terms and/or non-linear dielectric nature that allow the phonons to couple also gives rise to shifts in ν'. This much smaller effect (sometimes called the 'self-energy' shift) will shift ν' to lower frequencies on cooling. The 'self-energy' shift can only be separated from the pseudo-harmonicity shift if the spectra are taken under constant volume, pressure and temperature conditions, and since the pseudo-harmonicity shift usually predominates, ν' generally shifts to higher frequencies on cooling. In special cases

Width and temperature dependence

when the anharmonic forces are particularly strong (such as in ferroelectrics such as sodium titanate) 'self-energy' shifts may predominate.

Various theoretical studies of the temperature dependence of phonon frequencies have been made. Maradudin and Fein [446] have studied the effects of anharmonicity on neutron scattering, and their results were further confirmed by explicit expressions for thermal deformations and frequency shifts in terms of force constants obtained by Maradudin [445]. The results of these neutron scattering studies can be applied to infrared and Raman studies. Cowley [372, 448] and Ipatova, Maradudin and Wallis [447] have given more general treatments. Recently [455] the introduction of exact finite-strain parameters instead of the approximate parameters used above, together with the consideration of internal strains caused by the relative displacement of sublattices, has led to more exact theoretical expressions.

Mitra [12] and Loudon [5] have reviewed a number of experimental studies. $Mg(OH)_2$ is of interest [456] because of its unexpectedly large positive temperature shift due to the large anisotropic thermal expansion. Solid N_2 [457] and solid normal and deuterated ethane, ethylene, and acetylene [458] provide examples of temperature shifts in molecular crystals. Fig. 5.1 illustrates the variation of the width of ν' with temperature for silver bromide.

In some cases a variation of temperature will cause a phase change in the crystal and a particular phonon may have its force constant lessened in the phase change. Such 'soft' phonons have been the subject of a number of studies [55, 459–62].

5.3 MULTIPHONON PROCESSES

In the previous section it was seen that phonon phonon interaction was caused by the presence of anharmonic terms in the crystal potential and/or a non-linear dielectric nature, thus allowing the mechanism of these multiphonon processes to be understood. In general the $(n+1)$th order term in the potential energy (see (5.1)) and/or the nth order term in the dipole moment (see (5.2)) allow n-phonon transitions to occur.

5.3.1 *The infrared mechanism*

First consider the case of a crystal that has infrared active vibrations. When such a case was discussed in §4.6 it was seen that if the rate of exchange of energy between the phonon and photon of the polariton

Second order vibrational features

was slow compared to the transition probability for phonon–phonon interaction a combination band rather than a fundamental would occur. Fig. 5.2(a) illustrates the situation for a one phonon fundamental, where the polariton is real. Fig. 5.2(b) and (c) illustrate the situation for a two-phonon combination, where the polariton (in the polariton frequency range between o and ν_b (fig. 4.6)) is virtual and therefore not formed because of the fast phonon–phonon interaction. The detailed theory for

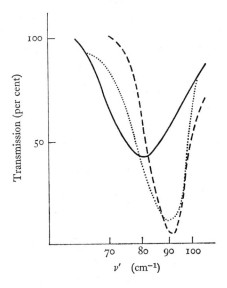

Fig. 5.1. Transmission spectrum of a thin film of silver bromide at various temperatures [463]. —— = 300 °K; .... = 100 °K; - - - - - = 4 °K.

the two phonon processes arising from the anharmonic mechanism has been given [442].

In the case of a crystal that has no infrared active vibrations processes of the type illustrated in fig. 5.2(a) to (c) are not possible. It is not possible for the photon to interact directly with only the anharmonic part of the potential energy and induce multiphonon processes [442], but direct interaction can occur in such cases via the second and higher order terms in the dipole moment (see (5.2)) that lead to a non-linear dielectric nature. The detailed theory for the two-phonon processes arising from the second order moment has been given [464] and fig. 5.2(d) and (e) illustrate such processes.

Multiphonon processes

5.3.2 The Raman mechanism

The virtual polariton of the Raman process (§4.7) can interact with phonons (H'_{EL}) in multiphonon processes through the presence of anharmonic forces and/or a non-linear dielectric nature. Such multiphonon processes known as *multiorder continuous spectra* are analogous to the infrared processes illustrated in fig. 5.2, and fig. 5.3(a) and (b) illustrate the mechanism [5, 465] for such Raman processes.

Another multiphonon process, known as the *second order line spectrum*, which is peculiar to the Raman effect, is also possible. In this process the

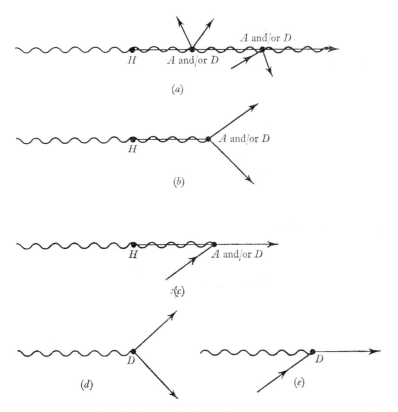

Fig. 5.2. Diagrams showing harmonic interactions (H) with photons leading to polaritons, and anharmonic (A) and non-linear dielectric (D) interactions leading to multiphonon (shown as two-phonon) processes. (a) Shows a fundamental, with polariton energy loss by multiphonon processes, (b) shows a combination, and (c) a difference multiphonon processes where the polariton is virtual, (d) and (e) show direct interaction of photons with phonons via D in combination and difference processes.

Second order vibrational features

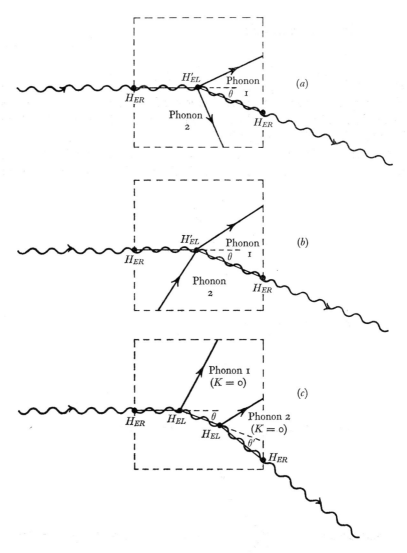

Fig. 5.3. The Raman scattering process showing interation vertices H_{ER} and H'_{EL} (which occurs via A and/or D) for multiphonon processes. (a) Is a second order continuous combination and (b) difference. (c) Is a second order line combination.

scattered photon of the first order process is rescattered within the crystal volume (illustrated in fig. 5.3(c)). The process increases in efficiency as the volume of the crystal increases, though it is, of course, essential that the first order Raman scattering for the two $K = 0$ phonons individually

Multiphonon processes

is allowed. Since the process is restricted to $K = 0$ phonons the bands are sharp giving rise to a line spectrum, and the spectra can be analyzed by the $K = 0$ selection rules discussed in §3.9.

5.3.3 Conservation of energy and wave vector

The processes discussed above are subject to the conservation of energy and wave vector. The conservation of energy requires

$$h\nu = \sum_i \pm h\nu_i \qquad (5.8)$$

in infrared studies, where ν is the photon frequency and ν_i the phonon frequency, and

$$h\nu = h\nu_f + \sum_i \pm h\nu_i \qquad (5.9)$$

in Raman studies. The conservation of wave vector requires

$$k = \sum_i \pm K_i + nG \qquad (5.10)$$

in infrared studies, where k is the photon wave vector and K_i is the phonon wave vector and G a reciprocal lattice vector (§2.8), and since in infrared studies $k \approx 0$

$$\sum_i \pm K_i = nG \qquad (5.11)$$

and

$$k_i = k_f + \sum_i \pm K_i + nG. \qquad (5.12)$$

For one- and two-phonon processes $n = 0$. This is clearly the case since the only way for two vectors to equal zero is for them to lie along the same direction in space. If they do this the lattice vector G must be zero (fig. 5.4). When a three-phonon process is concerned $n = 0$ or ∓ 1, because three vectors can be made equal to zero in ways that may or may not involve G (fig. 5.4 provides examples of these ways). Processes for which $n = 1$ are known as Umklapp process and are rare in spectroscopic studies [358], though Umklapp processes in phonon–phonon interactions cause the loss of crystal momentum that explains thermal conductivity [466, 24]. It is physically clear why G is not greater than 1, since if it was, one of the phonon wave vectors K would have a value greater than that of the first Brillouin zone.

5.3.4 Types of multiphonon process

Multiphonon processes can correspond to combinations, overtones, and differences:

(i) One phonon = fundamental.

Second order vibrational features

(ii) Two phonons = (a) both modes belong to the same irreducible representation of the same space group
= overtones,

(b) both modes belong to different irreducible representations
= combination (addition) if $K_1 = -K_2$
= combination (difference) if $K_1 = K_2$.

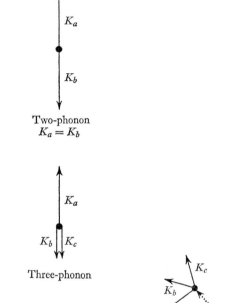

Fig. 5.4. Diagrams illustrating the conservation of wave vector in multiphonon processes. Solid lines represent phonons, dotted lines phonon vector constructional lines, and dashed lines represent the lattice vector G.

(iii) Three phonons = (a) pure combination,
(b) pure overtone, or
(c) pure mixture.

The process whereby anharmonicity allows multiphonon changes, and thus overtones and combinations is analogous to the gas-phase case [234, 467]. Repulsive anharmonic forces can cause binding between two-

Multiphonon processes

phonon states [468] leading to a two-phonon bound state with a frequency higher than twice the maximum single-phonon frequency. Combination bands may involve any combination of internal modes, external modes, or mixtures of internal and external modes.

5.3.5 The number of multiphonon bands

It follows from (5.11) and (5.12) that combination bands can occur with wave vector values throughout the Brillouin zone [469]. For a crystal containing N atoms *there may be as many as $3N$ modes capable of taking part in combination bands*, though there are only $3n$ (n = number of atoms in the smallest volume unit cell) that can act as fundamentals (§2.5). In practice, however, there are less than $3N$ modes capable of taking part in combination bands because of singularities in the phonon frequency distribution. These singularities occur where the dispersion curves for the individual branches are flat, which is the case at or near the *critical points* in the Brillouin zone. Critical points [30, 470, 471] are characterized by regions where there is a high density of phonon states per unit wave vector interval (and thus the dispersion curves are flat). For combination bands the combined density of states [472, 5] spectrum is used. This is obtained by counting the number of pairs of frequencies that occur in a given frequency interval, and fig. 5.5 illustrates the results of such a calculation for CaF_2 [473]. Most critical points are found at $K = 0$ and at specific points on the surface of the Brillouin zone in directions of high symmetry. In general for a three-dimensional crystal there are at least two critical points [470, 471], and in many cases, especially when the crystal symmetry is not high, the number of critical points will be much larger. Most of the critical points occur at points of high symmetry in the Brillouin zone and at $K = 0$. In the case of a face-centred cubic structure, for example, Phillips [474] has shown that critical points occur at the symmetry points of the Brillouin zone Γ, X, L and W, illustrated in fig. 5.6 (the definition and shapes of other Brillouin zones can be found discussed by Bradley and Cracknell [248]), the remaining critical points occurring on lines or planes of symmetry or at general points in the zone. The infrared spectra of $Mg(OH)_2$ [6], $Ca(OH)_2$ [293] and YF_3 [475] and the Raman spectrum of ZnS [476] and MgO [476a] provide examples of experimental studies where the multiphonon scattering at these critical points has been considered. In general if there are C critical points in the Brillouin zone of a crystal $3nC$ individual frequencies can take part in combination bands.

Second order vibrational features

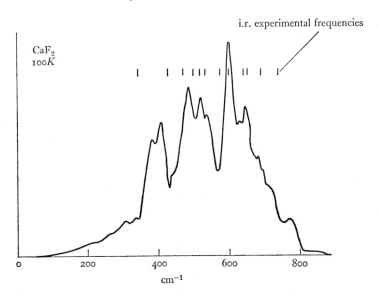

Fig. 5.5. Combined density of states spectrum calculated for CaF$_2$, compared with the experimentally observed infrared two-phonon bands. Reproduced by permission from Denham, Field, Morse and Wilkinson [473].

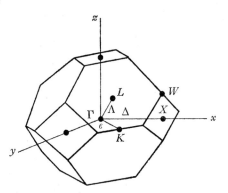

Fig. 5.6. Symmetry points of the first Brillouin zone of a face-centred cubic lattice.

5.3.6 *The activities of multiphonon bands*

The origin of the excitations involved in multiphonon bands can be understood by considering a classical approach to the activities of the bands. For a $K = 0$ fundamental vibration fig. 5.7(a) shows that there will be a dipole moment change for the entire crystal, if there is a dipole

146

Multiphonon processes

moment change for one molecule. If, however, $K \neq 0$, fig. 5.7(b) shows that even though there is a dipole moment change for a molecule (the same argument applies to polarizability changes when considering the Raman spectrum), the net dipole moment change for the whole crystal will cancel to zero. This is in agreement with the conservation of wave

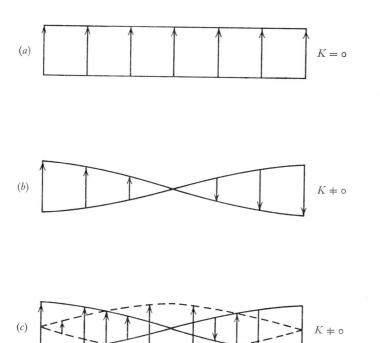

Fig. 5.7. Classical picture of phonons.

vector for fundamentals ($k_{\text{photon}} = K_{\text{phonon}}$). For two-phonon combinations or overtones, two phonons are involved, and these will be travelling in different directions. Two cases arise:

(i) The two phonons are superimposable (and thus appear as fig. 5.7(b)), and except when $K = 0$, there will always be a net zero dipole moment change for the whole crystal. In this case all the excitations are located on the same molecule.

(ii) The two phonons are not superimposable (and thus appear as fig. 5.7(c)). In this case the excitations are never located on the same molecule. It is now not possible to predict whether the combination of

Second order vibrational features

the two fundamentals will give rise to a dipole moment change for the whole crystal. The dipole moment changes for the fundamentals must sum to zero, though it is clearly not a physical necessity that the dipole moment changes for the combination sum to zero, since now these dipole moment combinations do not in general follow the phase determining action of a phonon wave. These modes cannot appear with any intensity unless the potential energy or dipole moment expression contains cross terms between neighbouring molecules or atoms. Selection rules can be derived that supplement the group theoretical selection rules by specifying conditions under which interactions between nearest-neighbour particles vanish. Anharmonic interactions are considered in cases where such interactions principally govern multiphonon process [476], and the non-linear dielectric nature is considered [477] where this is the principal multiphonon mechanism.

The classical approach, as illustrated in fig. 5.7, while satisfactory for the explanation of the activity of fundamentals and combinations of type (i) in terms of wave vector values, says nothing about which $K = 0$ combinations are active, and throws very little light on the $K \neq 0$ combination case (type (ii)). A proper analysis of multiphonon activities using group theory was given in §3.9, where it was seen that the vast majority of multiphonon bands will be Raman and infrared active for all directions of polarized light [478] or crystal orientations. Since each of the critical point phonons will be related to one of the irreducible representations of the crystal space group the application of space group theory can lead to multiphonon selection rules. These selection rules have so far only been derived for highly symmetrical crystals (examples have been given in §3.9), and in general for more complex crystals with large numbers of critical points most multiphonon bands will be infrared and Raman active.

5.3.7 *The temperature dependence of multiphonon bands*

The same factors which determine the width and temperature dependence of fundamentals (§5.2) will apply to multiphonon bands, though the greater number of phonons involved would lead to greater width due to increased phonon–phonon interaction possibilities, the observed width reflecting the combined width of the parent phonons. In addition it is necessary to consider whether the probability of a particular multiphonon process will also be temperature dependent. It has already been pointed out (§4.6) that multiphonon processes become more favoured at higher temperatures because the presence of more thermal phonons in the crystal leads to increased phonon–phonon interaction, and this will now be put on a more quantitative basis.

The average number of phonons in a lattice vibration of frequency ν (i.e. the number of quanta $h\nu$ sometimes called the phonon occupancy) is given by [479]

$$\bar{n} = \frac{1}{e^{-h\nu/kT} - 1}. \qquad (5.13)$$

Consider a transition from the state $\psi(\bar{n})$ to the state $\psi(\bar{n}+1)$ which will involve the creation of phonons. The matrix element for the calculation of transition probability can be written [466]

$$\langle \psi(\bar{n}) | \mathbf{a}^+ | \psi(\bar{n}+1) \rangle = (\bar{n}+1)^{\frac{1}{2}}, \qquad (5.14)$$

where $\mathbf{a}^+$ is the creation operator. Likewise for a transition from the state $\psi(\bar{n})$ to the state $\psi(\bar{n}-1)$ which will involve the destruction of phonons the matrix element can be written

$$\langle \psi(\bar{n}) | \mathbf{a}^- | \psi(\bar{n}-1) \rangle = (\bar{n})^{\frac{1}{2}}, \qquad (5.15)$$

where $\mathbf{a}^-$ is the annihilation operator. The infrared process for a one-phonon fundamental can be considered in terms of the absorption of a photon and emission of a phonon a process which will have a probability (from (5.14)) of $(\bar{n}+1)$. The overall absorption process must take into account the probability that a phonon will be destroyed and a photon emitted which will be (from (5.15)) $\bar{n}$. The overall probability for a one phonon fundamental will be given by $(\bar{n}+1) - \bar{n} = 1$ and is therefore independent of the number of quanta present in the crystal. This model is consistent with the polariton description if one considers the exchange of energy between the phonon and photon in terms of the virtual processes involving the destruction and creation of phonons and photons discussed above. When a two-phonon combination band is considered the overall probability is no longer independent of the number of quanta since the overall probability will likewise be given by

$$(\bar{n}_a+1)(\bar{n}_b+1) - \bar{n}_a\bar{n}_b = 1 + \bar{n}_a + \bar{n}_b$$

(when a and b refer to the two phonons K_a and K_b), leading to a temperature dependence of the probability given by (5.13). This amounts to saying that the rate of exchange of energy between the phonon and photon of the polariton becomes slow compared to the transition probability for phonon–phonon interaction which increases with the number of phonons given by $\bar{n}$ (when T is large

Second order vibrational features

(5.13) becomes $\bar{n} = kT/h\nu$). In Raman studies the probability will depend upon $\bar{n}$, even for fundamentals, since in these studies only the creation (Stokes line) and destruction (anti-Stokes line) of phonons is considered. This leads to a $(\bar{n}+1)$ (Stokes) and $\bar{n}$ (anti-Stokes) dependence for fundamentals and a $(1+\bar{n}_a+\bar{n}_b+\bar{n}_a\bar{n}_b)$ (Stokes) and $(\bar{n}_a\bar{n}_b)$ (anti-Stokes) dependence for two-phonon combinations and so on. Table 5.1 lists the temperature dependence factors for the various two- and three-phonon processes.

TABLE 5.1 *Temperature dependence factors for multiphonon processes*

Process	Infrared	Raman Stokes	Raman Anti-Stokes
$K_a+K_a=0$	$1+2\bar{n}_a$	$1+2\bar{n}_a+\bar{n}_a^2$	$\bar{n}_a^2$
$K_a+K_b=0$	$1+\bar{n}_a+\bar{n}_b$	$1+\bar{n}_a+\bar{n}_b+\bar{n}_a\bar{n}_b$	$\bar{n}_a\bar{n}_b$
$K_a-K_b=0$	$\bar{n}_b-\bar{n}_a$	$\bar{n}_a\bar{n}_b+\bar{n}_b$	$\bar{n}_a\bar{n}_b+\bar{n}_a$
$K_a+K_b+K_c=0, \pm G$	$1+\bar{n}_a+\bar{n}_b+\bar{n}_c$ $+\bar{n}_a\bar{n}_b+\bar{n}_b\bar{n}_c$ $+\bar{n}_a\bar{n}_c$	$1+\bar{n}_a+\bar{n}_b+\bar{n}_c$ $+\bar{n}_a\bar{n}_b+\bar{n}_b\bar{n}_c$ $+\bar{n}_a\bar{n}_c+\bar{n}_a\bar{n}_b\bar{n}_c$	$\bar{n}_a\bar{n}_b\bar{n}_c$
$K_a+K_b-K_c=0, \pm G$	$\bar{n}_c-\bar{n}_a\bar{n}_b+\bar{n}_a\bar{n}_c$ $+\bar{n}_b\bar{n}_c$	$\bar{n}_c+\bar{n}_a\bar{n}_b\bar{n}_c+\bar{n}_a\bar{n}_c$ $+\bar{n}_b\bar{n}_c$	$\bar{n}_a\bar{n}_b+\bar{n}_a\bar{n}_b\bar{n}_b$
$K_a-K_b-K_c=0, \pm G$	$\bar{n}_b\bar{n}_c-\bar{n}_a-\bar{n}_a\bar{n}_b$ $-\bar{n}_a\bar{n}_c$	$\bar{n}_b\bar{n}_c+\bar{n}_a\bar{n}_b\bar{n}_c$	$\bar{n}_a+\bar{n}_a\bar{n}_b\bar{n}_c$ $+\bar{n}_a\bar{n}_b+\bar{n}_a\bar{n}_c$

5.3.8 Multiphonon–single phonon mixing

Coupling can occur between multiphonon modes and single-phonon fundamental modes. This coupling exhibits the non-crossing (fig. 4.12) characteristics of resonance between two coupled oscillators and is analogous to Fermi-resonance in gas-phase systems. It is well illustrated in systems that possess a 'soft' single phonon mode that moves across the multiphonon mode, such as the coupling that occurs between a 'soft' zone centre optical phonon and two oppositely directed zone-edge acoustic phonons in quartz [459]. The situation has been analyzed theoretically [480] for two oppositely directed zone-edge acoustic phonons and a single-phonon mode and the coupled hybrid analyzed for different amounts of coupling in the first order Raman spectrum. A coupled hybrid between a one-phonon state and a number of two-phonon states with overlapping energy has been observed [481] in Raman studies leading to a strange series of line shapes.

5.3.9 *Examples of multiphonon processes*

Experimental studies of multiphonon processes have been reviewed by Mitra [12] and by Finch et al. [13].

5.4 BAND SHAPES

In this section the band shapes that are observed in experimental studies will be discussed. The width of fundamentals was discussed in §5.2 and §5.18, but in experimental studies additional features may contribute to the observed width.

(i) For a molecular case, each molecule can vibrate with hot bands characterized by vibrational transitions 1 to 2, 2 to 3, etc. In this case [482] the shift and broadening of spectral lines with increasing temperature is explained as being due to the increasing importance of hot fundamental bands originating from thermally excited states of the crystal. The broadening of the spectral line results because anharmonic forces cause the hot bands to occur at frequencies different from those of the fundamentals. At sufficiently high temperatures hot fundamentals will cause a frequency shift.

(ii) Unresolved solid state splittings will contribute to the observed width. These splittings may be the splittings of one fundamental of the infinite crystal by finite size effects (§4.10), and/or a number of unresolved fundamentals that occur at approximately the same frequency (such as correlation field components), and/or unresolved ordinary–extraordinary ray splitting in uniaxial and biaxial crystals (§4.9), particularly in powder studies.

(iii) Multiphonon–single phonon mixing (§5.3).

(iv) In infrared transmission studies, the superposition of reflection (and emission) spectra will contribute to the observed width. Reflection maxima have been seen (§4.4) to occur in the vicinity of absorption maxima, and will be strong for a strong absorption band (§4.6). It has also been seen (see (4.53)) that the reflection band will often be broad, and thus strong infrared transmission bands may often be broad through the superposition of a strong broad reflection maximum. The effect may be appreciable in powder samples [483, 484].

(v) In infrared transmission studies particle scattering and sample interference can also contribute to the observed width (these topics will be discussed in §5.7).

Second order vibrational features

(vi) Coupling of phonons to magnons and excitons (§ 6.4) may cause a reduction of linewidth [484a].

Even allowing for these various effects the widths of bands in solid state spectra are frequently not explained especially in more complicated systems. The effect of the superimposition of multiphonon bands has been considered as an additional contributer to band widths, and an approximate method discussed [485] for evaluating their contribution. All the factors that contribute to band shapes discussed above (apart from multiphonon bands) are expected to give rise to bands of a certain finite, though similar, width, in the absence of large differences in intensity (which would cause differences in the reflection contribution).

It has been seen that in all but the simplest and most symmetrical crystals the vast majority of the $3N$ multiphonon bands will be infrared and Raman active for all directions of polarized light or crystal orientation. Thus a very large number of superimposed combination and overtone bands would be expected for many infrared and Raman experiments. The intensity of these bands individually would be expected to be small (except in crystals with a large amount of anharmonic character (such as ferroelectric crystals [481]) where the intensity may be appreciable), but an appreciable intensity may be expected if a large number of such transitions overlap at a similar frequency.

Fig. 5.8 shows a typical dispersion relationship, modified to take account of different crystal orientations by showing each mode as a band rather than a line. In studies of powders, a random orientation of crystals will result, and the dispersion relation for one symmetry direction must be replaced by a superimposed diagram for all symmetry directions giving rise to the bands shown. A typical fundamental ($K = 0$) vibration is shown as $a-b$, and a typical $K \neq 0$ vibration as $c-d$ to $c-e$ (depending upon crystal orientation). Most active combination bands have been seen to be derived from $K \neq 0$ vibrations, but only $K = 0$ vibrations are fundamentals, and thus only if the $K = 0$ vibration has the same frequency $(a-b)$ as the $K \neq 0$ vibration $(c-d$ to $c-e)$ will the $K \neq 0$ combination be calculable from addition or subtraction of experimentally determined fundamental ($K = 0$) frequencies. The extent to which such an approximation is valid will thus depend upon the amount of dispersion in the dispersion relation. There would be expected to be little dispersion for covalently bonded molecules in molecular crystals (where the strong intramolecular forces are much greater than the weak intermolecular forces), and the dispersion would be expected to increase with

Band shapes

coupling between the molecules. Thus if a complete infrared or Raman spectrum is run and assuming that all the active fundamentals have been observed, a list of all possible combination, difference, and overtone bands (which will be for $K = 0$) can be calculated. By combining $K = 0$ combinations for all modes therefore, it is possible to get an *idea* of the *range* of all possible combinations, and the approximation will work best for covalent substances. In Raman spectra the absence of a reflection spectra will give better $K = 0$ fundamental data, and thus better combination band values.

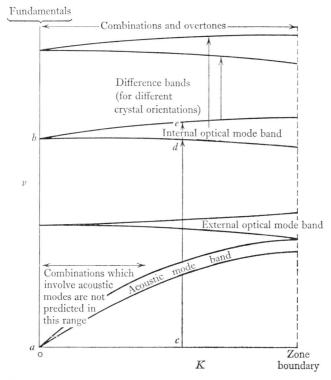

Fig. 5.8. General dispersion relation with bands to show different crystal orientations. Reproduced by permission from Sherwood [485].

The important qualitative results of this type of analysis are:

(i) Areas where there are a large number of combination bands would be expected to give broad bands and vice versa. Fig. 5.9 shows the treatment of experimental results from a complete infrared spectrum of a powder sample of iodine pentoxide [198]. This compound is a good example since it has a large number of absorption bands and these bands

Second order vibrational features

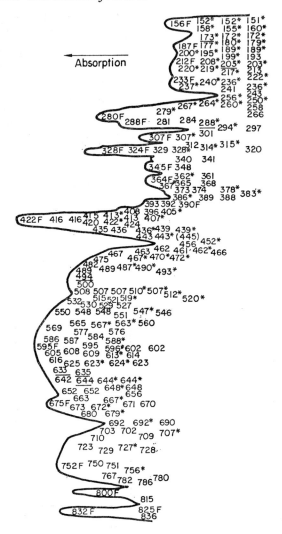

Fig. 5.9. Band profile of iodine pentoxide. * = difference band; F = fundamental band. Reproduced by permission from Sherwood [485].

are of medium intensity (reducing complications due to the reflection contribution), and the Raman frequencies have similar values to the infrared frequencies, and it is a covalent substance. The observed spectrum is shown as a solid line, and the numbers represent all the possible multiphonon bands (only two-phonon combination bands are shown for clarity, three-phonon bands will be of lesser intensity).

Band shapes

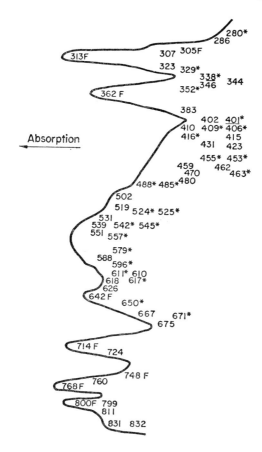

Fig. 5.10. Band profile of iodic acid * = difference band; F = fundamental band. Reproduced by permission from Sherwood [485].

(ii) Areas where there are a large number of combination bands but no fundamentals, may give an intense band due to a 'combination maximum' (an example of this may be seen in the solid CO spectrum [468]). The detection of possible 'combination maxima' may be important to avoid assignment as a fundamental. Fig. 5.10 shows the treatment of experimental results from a complete infrared spectrum of a powder sample of iodic acid [198] (a good example for the same reasons as iodine pentoxide). The same layout as fig. 5.9 is used. A 'combination maximum' can be seen at 577 cm^{-1}, being identified as such from the absence of a band at this value in the Raman spectrum (where all infrared bands are also Raman active as fundamentals).

155

Second order vibrational features

(iii) Identification of difference bands (shown with an asterisk in figs. 5.9 and 5.10) gives a useful qualitative idea of the effect of cooling to low temperatures, since these bands would be expected to disappear. Thus areas where there are a large number of difference bands would be expected to sharpen considerably. (This can be seen by comparing the band profile for iodine pentoxide (fig. 5.9) with the observed spectrum at room and liquid nitrogen temperature [198].)

5.5 INTERNAL MODES

5.5.1 *Shiftings*

In §4.3.2. it was seen how intermolecular forces cause a shift in frequency in going from the gas to the solid, given by the second term on the right-hand side of (4.50). The effect of environment on infrared absorption frequencies of substances in liquid solution has been extensively studied [487], and theories have been developed to deal with vibrational shifts, which should be capable of extension to the solid state. The shift of an internal mode may be expressed by [488]

$$\Delta \nu = \nu_{\text{solid}} - \nu_{\text{gas}}$$
$$= \Delta \nu_{\text{electrostatic}} + \Delta \nu_{\text{inductive}} + \Delta \nu_{\text{dispersive}} + \Delta \nu_{\text{repulsive}} \quad (5.16)$$

where the shift is given in terms of the sum of four terms representing electrostatic, inductive, dispersive, and repulsive contributions. The electrostatic contribution is a measure of the polar character of the molecules (the principal contribution in ionic compounds [489, 4]). The inductive contribution is caused by dipole-induced dipole interactions. The dispersive contribution is caused by London interactions that are present in all molecular crystals. The repulsive contribution is caused by the overlap of the electronic charge distributions of adjacent molecules in the crystal. Additional complications can be caused in solids by the presence of impurities that can interact with the molecules of the host and perturb its energy levels. Usually, however, this effect causes intermolecular interaction that results in the system being better studied as a complex [490]. Models for matrix isolated species have been discussed [488].

Shiftings can also be caused by the coupling of two or modes such as internal mode couplings or internal–external mode couplings and this will be discussed later.

5.5.2 *Splittings*

In §1.1 it was seen how the fundamental vibrational modes of a gas-phase molecule become split in the crystal due to static and correlation field

Internal modes

splittings, and examples of such splittings can be found in chapter 3 and elsewhere [491]. There has been a considerable amount of discussion about the relative magnitudes of site group and correlation field splitting. The relative magnitudes of the splittings depend upon the particular compound and the nature of the intermolecular forces. At first, site group splitting was considered to be greater than correlation field splitting [492, 493, 494], which appears to be the case for many ionic compounds (for example 25 cm^{-1} and 11 cm^{-1} for PH$_4^+$Br$^-$ [495]). Later it was noted that site group splitting was of the same order as correlation field splitting in covalent substances [489, 496] with van der Waals' intermolecular forces. Recently correlation field splittings have been observed in Raman studies (and calculated from repulsive forces between hydrogen atoms on neighbouring chains) of polymers where the splitting is very small [496 a].

Site group splitting and correlation field splittings can be distinguished by spectroscopic studies of dilute isotopic solid solutions. In such experiments [4] studies are made of solid solutions containing a low concentration of molecules which have one or more atoms, which take part in the internal vibration, isotopically substituted. The static field is the same for all the molecules whether isotopically substituted or not, but the isotopically substituted molecule will be largely unaffected by the vibrations of the surrounding molecules (even when the frequency difference is small), and thus will exhibit no correlation field splitting. The frequency observed for the isotopically substituted molecule will be that predicted for the oriented gas model (§4.3.2) though errors may arise because this model ignores intermolecular forces [357 a].

Splittings can also be caused by the coupling of two or more modes such as internal mode couplings or internal–external mode couplings and this will now be discussed.

5.5.3 *Coupling*

If the frequency of a vibration is close to that of another vibration then the assumption that each vibration occurs essentially uncoupled from its environment is inaccurate and breaks down. Coupling between the two vibrations must then be considered, and this will cause shiftings due to a splitting apart of the two mode frequencies and may lead to a splitting of any modes that are degenerate modes.

Second order vibrational features

Two types of coupling should be considered.

(i) Coupling between internal modes. This is analogous to coupling between modes in gas-phase systems.

(ii) Coupling between internal and external modes.

The two types of coupling can be considered physically in terms of the environment of the molecule executing an internal mode. In coupling of type (i), the molecular environment will be changing during the vibration, while in coupling of type (ii), the crystal environment (due to its translation or rotation as a whole in an external mode) will be changing during the vibration. In both cases the importance of the effect increases as the frequencies of the coupled modes approach one another, for example [497] for external modes below 300 cm^{-1} and an internal mode around 3000 cm^{-1} the coupling leads to a shift of only 0.2 cm^{-1}.

Even though the effect of coupling is small when the frequency difference is large, the possible small splitting of degenerate modes should be considered, since this splitting, if resolved, would add to the number of observed bands. In §3.8 it was seen how site group analysis could be used to determine whether degenerate molecular internal modes became non-degenerate in the crystal. Such an analysis can be applied to the case of an internal mode coupled to an external mode of much smaller frequency. In this case the site symmetry during the external mode (which will remain essentially constant during several cycles of the internal mode) may be lower than the equilibrium site symmetry which might lead to a splitting of a degenerate internal mode. Ammonium chloride provides an example [1] of such a case. The NH_4^+ ion lies at the centre of the cubic unit cell and has a site symmetry of T_d, and since there is only one molecule per unit cell there will be three degenerate translational external modes, three degenerate rotational external modes, and three degenerate acoustic modes. The external modes are illustrated in fig. 5.11 for the three modes along the x axis. In the case of the three degenerate translational external modes, which involve a displacement of the centre of gravity of the NH_4^+ ion, the site symmetry of this ion is reduced from T_d to C_{2v}. In the case of the three degenerate rotational external modes, which involve the rotation of the NH_4^+ ion about its centre of gravity, the site symmetry of this ion is reduced from T_d to C_2. Table 5.2 shows the effect that this lowering of site symmetry has on the degeneracy of the internal modes. In practice, the solid spectrum is composed of very broad bands, and resolution of the fine structure due to the predicted splitting cannot be observed [498, 499].

Internal modes

It is possible to predict whether internal mode degeneracies will be lifted by coupling with external modes without considering the site symmetries as described above. Hornig [1] has described how it is possible to apply an approximation analogous to the Born and Oppenheimer approximation (§2.3.2) to write down the wave functions for the internal and external modes which applies as long as the frequency

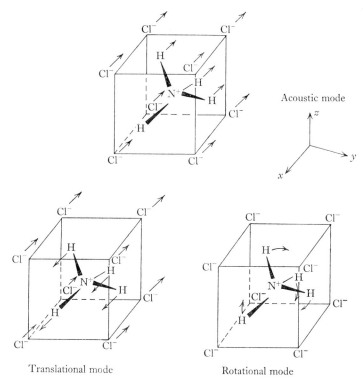

Fig. 5.11. Lattice vibrations in NH$_4$Cl.

difference between these two types of mode is large. The perturbation between these two types of mode is then considered, and it can be shown that a degenerate internal mode will be split by coupling with an external mode if the square of the irreducible representation of the degenerate internal mode (see the comment in §3.9.1 about evaluating this quantity) contains the irreducible representation of the external mode. It is assumed, however, that the internal mode is not affected by totally symmetric external modes.

Second order vibrational features

When the frequency difference between the internal and external modes becomes small, appreciable coupling can occur that leads to mixing of internal and external vibrations which belong to the same irreducible representations. This mixing has already been seen to cause problems in polymeric systems (§2.8.5), and has also been found to cause problems in salts with a fairly light, polyvalent, cation associated with a heavy complex anion.

TABLE 5.2 *Site group symmetry correlations for the internal modes of the NH_4^+ ion in solid NH_4Cl with and without internal–external mode coupling*

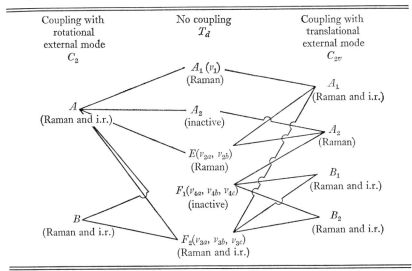

In some compounds the distinction between internal and external modes becomes of little value. Complex oxides of the type X_2YO_4 provide good examples of such compounds since it is often not possible to consider ionic bonding between X^{2+} and YO_4^{2-} ions and a three-dimensional array of X, Y, and O entities with varying degrees of ionic-covalent character. The vibrational spectra of such compounds whose modes belong to the same irreducible representation therefore exhibit appreciable degrees of mixing. The mixed modes can be distinguished from the other modes by studies of homogeneous families of pure spinels or by isomorphic replacement (which can be found reviewed elsewhere [500]) and by isotopic replacement. Isotopic replacement of the medium or heavy cations has recently been suggested and applied to the problem [501].

Internal modes

The isotopic replacement method provides a very clear idea of the presence of mixed modes for either the case just discussed or for cases of mixing of internal and external modes, even though isotopic shifts are small and the isotopes sometimes difficult to obtain. Table 5.3 provides examples of the method. The small, but significant isotopic shift for the ν_4 internal mode of $CaMoO_4$ indicates some mixing with a translational external mode (which would involve both Ca^{2+} and MoO_4^{2-} motion). The small isotopic shift for the ν_4 mode of Ni_2GeO_4 (a compound of the type X_2YO_4) for both Ni and Ge indicates a mixed vibration which is predominantly a Ni—O vibration, but contains some contribution from Ge—O vibrations.

TABLE 5.3 *Vibrations for isotopically substituted $CaMoO_4$ and Ni_2GeO_4 from the data of Tarte and Preudhomme* [501]

$CaMoO_4$ ν_4 internal mode	
$^{40}CaMoO_4 = 329$ cm^{-1}	$^{44}CaMoO_4 = 326$ cm^{-1}
Ni_2GeO_4 ν_4 mode	
$Ni_2{}^{70}GeO_4 = 201$ cm^{-1}	$Ni_2{}^{76}GeO_4 = 199.5$ cm^{-1}
$^{58}Ni_2GeO_4 = 201$ cm^{-1}	$^{62}Ni_2GeO_4 = 196.5$ cm^{-1}

5.5.4 *Internal rotation*

There have been many experimental and theoretical studies of the potential energy barrier to internal rotation in various organic molecules, and the subject has been reviewed [502, 503]. Torsional modes (not to be confused with external librational modes) involving such internal rotation have been studied by vibrational spectroscopy in order to try to evaluate this potential energy barrier. The difficulty with the study of such torsional modes by vibrational spectroscopy in the gaseous or liquid states is that these modes may be inactive. In solid state studies, however, such modes may become active because of the lowering of symmetry in the crystal (e.g. in studies of $M(CH_3)_4$ type molecules [504]).

5.5.5 *The effect of molecule deformation*

It has been assumed so far that molecules move as rigid bodies in internal modes. Intermolecular forces may distort the molecules raising the internal mode frequencies and giving them a K dependence. The lattice dynamics of naphthalene, which behaves in this way, has recently been investigated [505].

Second order vibrational features

5.6 EXTERNAL MODES

In §2.6 it was seen that there are two types of external mode, namely translational modes and rotational or librational modes. The frequency of a translational mode will depend upon the total mass of the translating body, and the frequency of a rotational mode will depend upon the moment of inertia for the rotation axis concerned. The two types of mode can therefore be identified by isotopic replacement. Thus isotopic substitution makes little effect on the frequencies of translational modes, but may make an appreciable effect on the frequency of a rotational mode if a rotating atom is substituted. It is, however, incorrect to suppose that a frequency shift on substitution indicates a rotational mode since translational modes will always give a small but finite frequency shift, but rotational modes will give no frequency shift if the rotation is about the substituted atom. Table 5.4 gives examples that illustrate the isotopic substitution technique.

TABLE 5.4 *External modes for isotopically substituted molecules which allow translational and rotational modes to be distinguished*

Molecule	Frequency of external mode (cm^{-1})	Isotopic frequency ratio	Comments	Reference
HCl	86	0.97	Translational mode (since there is little frequency change)	506
DCl	89			506
Ca^{40}MoO$_4$	237	1.058	Translational mode (since there is little frequency change)	501
Ca^{44}MoO$_4$	224			501
HCl	217	1.28	Rotational mode (since appreciable frequency change indicates that the rotating atom has been substituted)	506
DCl	169			506
Ca^{40}MoO$_4$	153	1.00	Rotational mode (since no frequency change indicates that the Ca atom is not involved which indicates either a rotational mode or a translational mode involving only the MoO$_4$ group)	501
Ca^{44}MoO$_4$	153			501

It was also seen in §2.5 however that the two types of external mode could not be clearly differentiated in the non-centrosymmetric case when mixing between the two types of mode occurs. In the centrosymmetric case mixing occurs when $K \neq 0$. This mixing has been demonstrated by Sandor [507] and Venkataraman and Sahni [47] by considering a

External modes

one-dimensional array of identical ring-type molecules each consisting of six close-packed spherical atoms. The translational acoustic mode (fig. 5.12) is purely translational at $K = 0$ and purely rotational at $K = \pi/3a$. The converse is true of the rotational optical mode. For general K values each mode is a mixed mode (this result is illustrated in fig. 5.13). Therefore nearly all external modes should be considered hybrids of translational and rotational modes though they are often predominantly of one type.

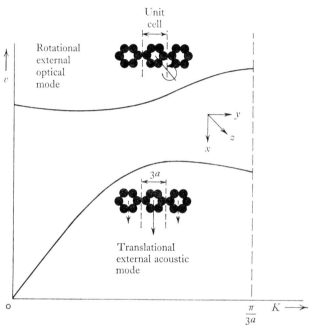

Fig. 5.12. Dispersion relation for rotational external optical mode and translational external acoustic mode for a one-dimensional array of identical ring-type molecules each consisting of six close-packed spherical atoms. The diagrams illustrate the molecular motion. The diagram is based on Sandor [507].

There have been a number of studies of predominantly rotational external modes, some involving combination band fine structure on internal modes [508, 509]. These studies have sometimes been directed at finding the potential energy barrier to free rotation [510, 511]. The quantum of rotational or librational external modes has sometimes been called a *libron* by analogy with the word phonon and the rotational external modes called librons. This nomenclature is inexact in view of the hybrid nature of such rotational external modes. The $K = 0$ rotational external modes

Second order vibrational features

in solid hydrogen are purely rotational and there has been appreciable interest in such librons [512, 513]. In particular, solid o-H_2 undergoes a phase change at 1.5° K from a f.c.c. lattice ($<$ 1.5° K) to a h.c.p. lattice ($>$ 1.5° K) and the energy of the librons provides corrections to the ground state energies in these forms [514, 515]. Libron–libron interactions have been discussed [516, 517], and the $K = 0$ libron spectrum has been observed by infrared [518, 255] and Raman studies [517, 519].

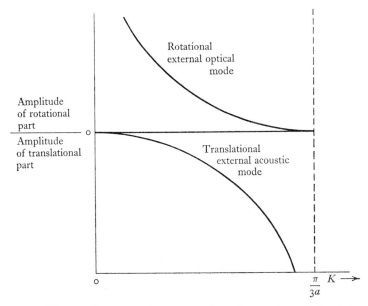

Fig. 5.13. Diagram illustrating the amount of rotational and translational character in the mixed mode for various K values. The diagram is based on Sandor [507].

Sometimes the potential energy barrier to free rotation is so small that free rotation can occur in the crystalline state [6]. Hydrogen is one of the few molecules that can perform free rotation in the solid; this has been detected by infrared and Raman studies [520–4]. Similar effects have been observed in HD [525]. By analogy with the word phonon such modes have been called *rotons*.

5.7 PARTICLE SCATTERING AND SAMPLE INTERFERENCE

The complications involved in the studies of powders have already been discussed in §4.10.5. In addition the powder particles themselves can cause elastic scattering of the incident radiation. This type of elastic

Particle scattering, sample interference

scattering is known as *Tyndall scattering*. In the elastic scattering described previously (§4.7) scattering events on a molecular level were considered. In Tyndall scattering scattering by a whole section of crystal making up a particle is considered. In this case the varying dipole moment, $\mathbf{M}$, given by (4.66) can be represented on a particle level by

$$|\mathbf{M}| = V|\mathbf{P}|\,|\mathbf{E}_i|, \tag{5.17}$$

where V is the volume of the particle and $\mathbf{E}_i$ is the macroscopic electric field inside the particle. The intensity of elastically scattered radiation is given by

$$I\text{ (elastic scattering)} = \frac{1}{8\pi}\left(\frac{|\mathbf{M}^2|}{|R^2|}\right)\frac{1}{\lambda^4}\sin^2\theta''', \tag{5.18}$$

where R is the distance from the source, and θ''' the angle from the polar direction from the incident radiation of wavelength λ. Therefore

$$I\text{ (Tyndall scattering)} = \frac{1}{8\pi}\left(\frac{V^2|\mathbf{P}|^2\,|\mathbf{E}_i|^2}{|R^2|}\right)\frac{1}{\lambda^4}\sin^2\theta'''. \tag{5.19}$$

Thus shorter wavelengths are scattered to a much greater extent than longer wavelengths. When the particle sizes are much larger than the wavelength of the incident radiation the large V^2 term in (5.19) will mean that there will be an appreciable intensity of scattering for various wavelengths (thus, for example, clouds appear white). The vibrational spectrum of large particles will thus have a uniform scattering background and no complications will result. When the particle sizes become comparable with or less than the wavelength of the incident radiation there will only be an appreciable intensity of scattering for shorter wavelengths (thus, for example, tobacco smoke appears blue). The vibrational spectrum of small particles will thus have a non-uniform scattering background and this scattering can shift or distort the transmission spectrum, and might even introduce scattering rather than absorption maxima. It should also be noted that the intensity of scattering will depend upon the shapes of the particles because $\mathbf{P}$ is shape dependent. $\mathbf{P}$ is given by (4.3) and (4.24) can be used to evaluate $\mathbf{E}_i$ in terms of $\mathbf{E}_r$, the field of the incident radiation. In this case, however, (4.24) is applied to the macroscopic particle case rather than the microscopic ionic case. $\mathbf{E}_2$ can be taken as zero, and $\mathbf{P}$ can be expressed as

$$\mathbf{P} = (\epsilon-1)(\mathbf{E}_r - \mathbf{E}_1). \tag{5.20}$$

$\mathbf{E}_1$ depends upon the shape of the particle leading to a shape dependence

Second order vibrational features

for P. For a spherical particle it is given by (4.25) and for transverse polarization of a plate-like particle it is given by

$$E_1 = 4\pi P. \qquad (5.21)$$

Thus it can be seen that $|P|$ and thus I would tend to infinity when

$$\epsilon = (1-(1/x)), \qquad (5.22)$$

where $x = \frac{1}{3}$ for a spherical sample and $x = 1$ for a plate-like particle. This shape dependence will obviously be important in cases of a non-uniform scattering background. Thus non-uniform scattering complications arise in powder samples particularly at longer wavelengths. Such complications have been well illustrated by experiments where the effect of variation of particle size and shape on the vibrational spectrum has been studied (e.g. in studies of MgO [526, 360] and MnO powders [360] and BaTiO$_3$, Al$_2$O$_3$, AlN, and BaO$_2$ powders [527]) notwithstanding the effects already discussed in §4.10.5.

Powder samples have been studied by spreading the powder between spectroscopic windows, or by techniques involving carrier liquids. These carrier liquids are volatile solvents such as ether, chloroform, petrol, and alcohol and the sample may be transferred via a glass plate [527].

Experimentally the amount of Tyndall scattering may be reduced by using a suspension of particles in a non-volatile liquid to give a mull, though the method is complicated by absorption due to the mull liquid. The mulling liquid may be dispensed with and the amount of Tyndall scattering reduced by compacting the particles in some window material. This method may, however, give erroneous results due to reaction with the window material [198].

In the case of samples of appreciable thickness (at least several wavelengths thick), additional complications can arise because of interference within the sample of the transmitted radiation. Such interference gives rise to periodic variation in the intensity of a given mode with sample thickness.

5.8 INFRARED INTENSITIES

In §3.9 it was seen that the absorption probability for infrared transitions contained the integral (3.70). The magnitude of this integral will therefore be directly proportional to the intensity of the mode. The intensity of the mode can be expressed in terms of an absorption coefficient A' defined by

$$I = I_0 e^{-A'x}, \qquad (5.23)$$

Infrared intensities

where I is the transmitted energy, I_0 the incident energy and x the sample thickness. For a band containing a number of superimposed modes A' can be replaced by A'' where

$$A'' = \int_{\text{band}} A' \, d\nu \tag{5.24}$$

and A'' represents the summation over the various components. A' and A'' can be shown to be related to the integral (3.70) by [528]:

$$A' = \frac{8\pi^3}{3hc} \nu \left\{ \int \psi_f^* \bar{\mu} \psi_i \, d\tau \right\}^2 (A_i - A_f) \left| \frac{E_l}{E} \right| \frac{1}{\bar{n}}, \tag{5.25}$$

$$A'' = \frac{8\pi^3}{3hc} \nu \sum_k \left\{ \int \psi_f^* \bar{\mu}_k \psi_i \, d\tau \right\}^2 (A_i - A_f) \left| \frac{E_l}{E} \right| \frac{1}{\bar{n}}, \tag{5.26}$$

where A_i and A_f refer to the number of ion pairs or molecules in a unit volume in the initial and final states. $|E_l/E| \, 1/\bar{n}$ is the field effect in the solid where [529]

$$\left| \frac{E_l}{E} \right| \frac{1}{\bar{n}} = \frac{A''_{\text{solid}}}{A'_{\text{gas}}} \tag{5.27}$$

and $\bar{n}$ is the average refractive index over the band given by [530, 531]

$$\bar{n} = \frac{\int_{\text{band}} nA' \, d\nu}{\int_{\text{band}} A' \, d\nu}. \tag{5.28}$$

Assuming the Boltzmann distribution law

$$(A_i - A_f) \approx A_i = A. \tag{5.29}$$

The integrals (3.70) in (5.25) and (5.26) must be evaluated. $\bar{\mu}$ refers to the average over the three directions x, y and z, i.e.

$$\bar{\mu} = \tfrac{1}{3}(\mu_x + \mu_y + \mu_z). \tag{5.30}$$

μ_ξ was given by (4.34) and refers to the instantaneous dipole moment of each ion pair of molecule for the phonon mode in the direction ξ (= x or y or z). It must be remembered that the solid state intensity is a tensor A''_ξ where

$$A'' = \tfrac{1}{3}(A''_x + A''_y + A''_z). \tag{5.31}$$

Thus the solid state intensity depends upon crystal position as found experimentally [532].

μ_ξ can be expressed as a series in terms of Q_k, and assuming no initial

Second order vibrational features

dipole moment (the higher terms are ignored since they give rise to multiphonon processes):

$$\mu_\xi = \sum_k \left(\frac{\partial \mu_\xi}{\partial Q_k}\right) Q_k + \tfrac{1}{2} \sum_k \left(\frac{\partial^2 \mu_\xi}{\partial Q_k^2}\right) Q_k^2 + \ldots \quad (5.32)$$

Since ψ_i and ψ_f are functions of the normal coordinates the integral (3.70) can be rewritten [533]

$$\int \psi_f^* \mu_\xi \psi_i \, d\tau = \left(\int \psi_f^* Q_k \psi_i \, dQ_k\right) \frac{\partial \mu_\xi}{\partial Q_k}, \quad (5.33)$$

$$= \left(\frac{h}{8\pi^2 c \nu_0''}\right)^{\tfrac{1}{2}} \frac{\partial \mu_\xi}{\partial Q_k}, \quad (5.34)$$

where ν_0'' is the zero order vibrational frequency. In going from (5.33) to (5.34) the harmonic oscillator approximation is assumed, and Burdett [534] has discussed the effect of a breakdown in this approximation. Substituting (5.34) and (5.29) in (5.26) using (5.31)

$$A_\xi'' = \frac{A\pi}{c^2} \frac{\nu}{\nu_0''} \sum_k \left(\frac{\partial \mu_\xi}{\partial Q_k}\right)^2 \left|\frac{E_l}{E}\right| \frac{1}{n}. \quad (5.35)$$

(5.35) may be written in the approximate form taking $(\nu/\nu_0'') \approx 1$

$$A_\xi'' = \frac{1}{n} \left|\frac{E_l}{E}\right| \frac{A\pi}{c^2} \sum_k \left(\frac{\partial \mu_\xi}{\partial Q_k}\right)^2. \quad (5.36)$$

For a degenerate mode the right-hand side of (5.36) is multiplied by the degeneracy.

(5.36) may also be derived from (4.47) by substituting for $(\epsilon_\nu - \epsilon_\infty)$ the relation of Schatz [530, 531] (where the factor $1/c$ has been introduced so that A_ξ'' is expressed in the same units as in (5.36)):

$$\epsilon_\nu - \epsilon_\infty = \frac{c^2}{\pi^2} \sum_k \frac{\bar{n} A_\xi''}{(\bar{\nu}_{kt})^2 - \nu^2}. \quad (5.37)$$

(5.36) gives A_ξ'' in cm per unit volume for $(\partial \mu_\xi/\partial Q_k)$ expressed in units of $cm^{\tfrac{3}{2}} sec^{-1}$. The usual units of A_ξ'' however are $cm\,mmol^{-1}$ where $1\,cm\,mmol^{-1}$ is a unit known as a *dark*. To obtain A_ξ'' in darks, A_ξ'' in cm per unit volume is divided by the density (ρ') in $mmol\,cm^{-3}$ (5.36) can thus be written for A_ξ'' in darks:

$$A_\xi'' \text{ (in darks)} = \frac{1}{n} \left|\frac{E_l}{E}\right| \frac{N_0 \pi}{1000 c^2} \sum_k \left(\frac{\partial \mu_\xi}{\partial Q_k}\right)^2, \quad (5.38)$$

where N_0 is the Avogadro number and A' is given by

$$I = I_0 e^{-A'\rho' x}, \qquad (5.39)$$

where x is in cm. Data is often [529, 535] presented in the form of a Beer–Lambert law plot where the integrated absorption

$$\int_{\text{band}} \log_e(I_0/I) \, d\nu$$

is plotted against x. Values of $\partial\mu_\xi/\partial Q_k$ are available [536, 533, 537], though in most cases only $\partial\mu/\partial Q_k$ for the gas phase is listed. It must be remembered that $\partial\mu_\xi/\partial Q_k$ is a tensor quantity depending upon the direction ξ of the phonon, and also that it does not apply to individual molecules or ion pairs but to the whole unit cell. Care must thus be taken before comparing solid and gas-phase values for this quantity, since $\partial\mu_\xi/\partial Q_k$ for the solid will (depending upon the degree of intermolecular interaction) approximate to the vector summation of the appropriate $\partial\mu/\partial Q_k$ for the individual molecules of the unit cell [538]. It should also be noted that $\mathbf{E}_l$ is a tensor and the field effect (see 5.27) can be evaluated using Buckingham's theory [539, 540]. $\mathbf{E}_l$ can be expressed by [541, 542]

$$\mathbf{E}_l = \mathbf{E} + \sum_{r \neq s} \mathbf{S}_{rs} \mathbf{\mu}_{\xi s}, \qquad (5.39a)$$

where $\mathbf{S}$ is the field propagation tensor [543] and $\mathbf{\mu}_{\xi s}$ the dipole moment in the crystal at the sth site (the unit cell dipole moment) for a collection of dipoles making up the crystal. $\mathbf{\mu}_\xi$ can be expressed as [544]

$$\mathbf{\mu}_\xi = (\mathbf{U} - \alpha_\xi \mathbf{S})^{-1}(\mathbf{\mu} + \alpha_\xi \mathbf{E}), \qquad (5.39b)$$

where $\mathbf{\mu}$ is the intrinsic dipole moment of the dipole (the gas-phase dipole moment), $\mathbf{U}$ a unit vector in the direction ξ, and α_ξ the polarizability tensor (§3.9). By writing down the interaction energy between the molecular dipoles and expressing in terms of the unit cell parameters $(\partial\mu_\xi/\partial Q_k)$ and the gas-phase $(\partial\mu/\partial Q_k)$ can be related [541, 542]

$$\frac{\partial\mu_\xi}{\partial Q_k} = (\mathbf{U} - \alpha_\xi \mathbf{S})^{-1} \left(\frac{\partial \alpha_\xi}{\partial Q_k} \mathbf{S}(\mathbf{U} - \alpha_\xi \mathbf{S})^{-1} \mathbf{\mu} + \frac{\partial \mu}{\partial Q_k} \right). \qquad (5.40)$$

$\mathbf{\mu}_\xi$ can arise from three sources.

(i) The permanent dipole moment in the molecules or ions in the crystal which gives rise to a large $(\partial\mu_\xi/\partial Q_k)$.

(ii) The induced dipole moment caused by the local field of the neighbouring molecular quadrupole and dipole moments.

Second order vibrational features

(iii) The dipole moment caused by short range forces that cause distortion of the electronic cloud.

These three sources of dipole moment contribution have been fully discussed by Schnepp [545].

5.8.1 Internal modes

In many cases there are differences between solid state and gas-phase intensities that are much greater than the difference in the frequencies. The intensity difference arises because of the intermolecular interactions in the solid discussed above. The differences are sometimes very striking e.g. the ν_1 mode in ClCN is about 900 darks in the gas phase but 0 darks in the solid [538]. The detailed study of solid state intensities thus provides useful information about the intermolecular interactions in the solid state though the problem is still not fully understood, and there are appreciable experimental difficulties. It is only recently that reliable absolute intensities have been obtained and these studies have been reviewed [538].

Intensity studies have been used to distinguish between longitudinal and transverse modes. The four bands seen in Raman studies of hydrogen and deuterium halide crystals have been assigned either to a four molecule primitive unit cell (which disagrees with diffraction studies) or to two transverse and two longitudinal modes. Using the dipolar coupling model used to derive (5.40) the longitudinal frequencies have been calculated from the infrared intensities giving excellent agreement with the Raman frequencies.

5.8.2 External modes

The detailed study of the intensities of external modes provides useful information about the electric moments μ_ξ. On the basis of these moments it is possible to make general observations about the intensities of the two types of external mode.

(i) Rotational modes. The intensity depends mainly upon the size of the permanent dipole moment of the rotating molecule. Thus for molecules with large dipole moments (e.g. hydrogen halides [506, 546], ClCN and BrCN [541] and OCS [547]) rotational external modes appear with appreciable intensity.

(ii) Translational modes. These modes would appear with no intensity if they involved the displacement of rigid unpolarizable atoms or molecules. In fact their intensity may result partly from mixing with rotational

Infrared intensities

modes of finite intensity and partly from polarization during displacement leading to an induced dipole moment. Thus the translational external modes in nitrogen [545, 118] appear with finite intensity because of the quadrupole-induced dipole moment. For charged ions in ionic compounds the dipole moment is of course very large and such modes appear with appreciable intensity.

Reliable absolute intensity measurements have included the studies of the rotational modes of OCS [547], halogens [548], HCl and HBr [548a], α-N$_2$ [549], and ClCN and BrCN [541] and the translational modes of ethylene [550] and CO$_2$ [551].

5.9 RAMAN INTENSITIES

In §3.9 it was seen that the probability of a Raman scattering event depended upon the integral (3.77). The magnitude of this integral will therefore be directly proportional to the intensity of the mode. The intensity of a Raman mode of frequency ν' for an exciting line of frequency ν is given by [552]

$$I = \frac{64\pi^2 A}{3c^2}(\nu-\nu')^4 \left\{ \int \psi_f^* \bar{\alpha} \psi_i \, d\tau \right\}^2 \mathbf{E}^2. \tag{5.41}$$

$\bar{\alpha}$ refers to the average value of α given by

$$\bar{\alpha} = \tfrac{1}{3}(\alpha_{xx}+\alpha_{yy}+\alpha_{zz}). \tag{5.42}$$

α_ξ was given by (3.71) and refers to the component of the polarizability tensor for the phonon mode in the direction ξ. As in infrared studies the Raman intensity will depend upon crystal position.

Using the polarizability theory of Placzek [553], α_ξ may be expressed as a series in terms of Q_k. This series will be identical to (5.32) with μ_ξ replaced by α_ξ. Rewriting the integral (3.71) as (5.34) with μ_ξ replaced by α_ξ the intensity can be written:

$$I_\xi = \frac{8A}{c^4\nu_0'}(\nu-\nu')^4 \left(\frac{\partial \alpha_\xi}{\partial Q_k}\right)^2 \mathbf{E}^2. \tag{5.43}$$

For a degenerate mode the right-hand side of (5.43) is multiplied by the degeneracy.

In the case of crystal powders or gases where there is a random orienta-

Second order vibrational features

tion of molecules the intensity is often expressed in terms of $\bar{\alpha}$, and γ'' the anisotropy. The anisotropy is given by

$$\gamma''^2 = \frac{1}{2}\left[\left(\frac{\partial\alpha_{xx}}{\partial Q_k} - \frac{\partial\alpha_{yy}}{\partial Q_k}\right)^2 + \left(\frac{\partial\alpha_{yy}}{\partial Q_k} - \frac{\partial\alpha_{zz}}{\partial Q_k}\right)^2 + \left(\frac{\partial\alpha_{zz}}{\partial Q_k} - \frac{\partial\alpha_{xx}}{\partial Q_k}\right)^2 \right.$$
$$\left. + 6\left(\left(\frac{\partial\alpha_{xy}}{\partial Q_k}\right)^2 + \left(\frac{\partial\alpha_{yz}}{\partial Q_k}\right)^2 + \left(\frac{\partial\alpha_{zx}}{\partial Q_k}\right)^2\right)\right], \quad (5.44)$$

and the intensity for a given value of θ (fig. 2.23) is [554]:

$$I_\theta = \frac{\text{Constant } I_0 A(\nu - \nu')^4}{\nu'(1 - e^{-h\nu'/kT})}$$
$$\times \left\{\left[45\left(\frac{\partial\alpha_\xi}{\partial Q_k}\right)^2 + 7\gamma''^2\right](1 + \cos^2\theta) + 6\gamma''^2 \sin^2\theta\right\}, \quad (5.45)$$

where the term $(1 - e^{-h\nu'/kT})$ takes vibrationally excited states into account, leading to a temperature dependence (experimental studies of which have been reviewed [5]).

When $\theta = 90°$ the polarization properties of the Raman scattered light may be considered. For non-polarized incident light characterized by the two electric vectors $\mathbf{E}_x$ and $\mathbf{E}_y$ of equal magnitude, the 90° scattered light will have intensities I_y and I_z (fig. 5.14). In general $I_z \neq I_y$, but in the special case of highly symmetrical systems with a spherical polarizability ellipsoid $I_z = 0$. A quantity ρ'', the degree of depolarization, may be defined

$$\rho'' = I_z/I_y, \quad (5.46)$$

which is clearly zero for completely polarized light.

In single crystal studies the scattering efficiency S is sometimes considered instead of I_ξ. S is defined [555] as

$$S = N(\nu_f)/N(\nu), \quad (5.47)$$

where $N(\nu_f)$ is the number of scattered photons of frequency ν_f produced per unit time per unit cross-sectional area in the solid angle $d\Omega$ about the direction of observation, and $N(\nu)$ is the number of incident photons of frequency ν per unit time per unit cross-sectional area. For $\theta = 90°$ and unpolarized incident radiation S is given [556] by

$$S_\xi = \frac{3h\nu^4 L \, d\Omega}{2\pi\rho c^4 \nu'}\left(\frac{\partial\alpha_\xi}{\partial Q_k}\right)^2 \frac{1}{(1 - e^{-h\nu'/kT})}, \quad (5.48)$$

where ρ is the crystal density and L the length of the crystal in the direction ξ. ρ'' can be expressed in terms of $S_\parallel$ (light polarized in the plane of

Raman intensities

scattering, the xz plane) and $S_\perp$ (light polarized perpendicular to the plane of scattering, the xy plane)

$$\rho'' = \frac{S_\parallel}{S_\perp}. \tag{5.49}$$

More detailed theoretical treatments of scattering efficiencies can be found in the works of Born and Huang [556], Loudon [5] and Cowley [557]. Loudon [5] has reviewed the situation in piezoelectric crystals where differences in intensities between longitudinal optical and transverse optical phonons arise because of the effect of the electric field associated with the longitudinal optical phonon [558].

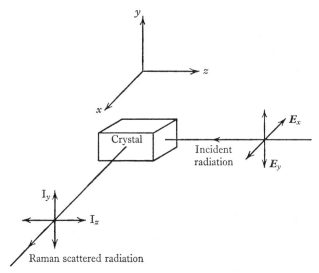

Fig. 5.14. Polarization properties of Raman scattered light for $\theta = 90°$.

The evaluation of $\partial \alpha / \partial Q_k$ values from gas and solution studies has been reviewed by Hester [554] and Murphy, Holzer and Bernstein [559] list the values obtained in mercury arc studies. Many of the errors involved in arc studies (such convergence errors, reflection losses, and spectral sensitivity of the photomultiplier tube) can be eliminated by using laser excitation.

Recent developments in the theories of Raman intensities have been reviewed by Tang and Albrecht [560].

The experimental studies of the Raman intensities of external modes have included work on oxygen [299], halogens [561], anthracene and naphthalene [562].

Second order vibrational features

5.10 RESONANT RAMAN SCATTERING

In §4.7 it was seen that the frequency of the monochromatic electromagnetic radiation ν lay in the range of exciton transitions leading to the consideration of the virtual intermediate state as a polariton (of the photon–exciton type). Normally $\nu \neq \nu_0^e$ (where ν_0^e is the exciton transition frequency). In the *resonance Raman effect* $\nu \approx \nu_0^e$ and it can be seen from (4.66) that such a situation leads to appreciable elastic and inelastic scattering as $|M|$ tends to infinity. The greatly increased amount of scattered radiation may, however, be appreciably absorbed leading to exciton transitions, and there may be complications due to superimposed fluorescence emission. The exciton transitions may be accompanied by sample photolysis with subsequent sample destruction. Despite these problems resonance Raman studies have often led to appreciable increases in intensity.

The increase in the scattering efficiency S is predicted by the polariton descriptions (§4.7) and confirmed by experimental studies [465, 563–6].

In resonant Raman scattering the usual selection rules no longer apply, and the reasons for the breakdown of these selection rules have been examined [567, 568]. The theory has been discussed [568a].

Resonance Raman scattering in liquids has been fully reviewed [569].

5.11 BRILLOUIN SCATTERING

In §4.8 it was seen how acoustic phonons could be inelastically scattered because of the finite size of K in Raman studies. The very small Brillouin shifts ($\nu \pm \nu_f$ is only 1–5 cm^{-1}) mean that Brillouin scattered bands occur very close to the Rayleigh scattered exciting line and high resolution techniques must be used.

Born and Huang [34] have described the theory of Brillouin scattering for the macroscopic model and give expressions for the scattering efficiency (note the error in these expressions pointed out by Loudon [5]). Other models have been reviewed [5].

The Brillouin scattering spectrum allows the acoustic phonon velocities to be determined from

$$\frac{\nu - \nu_f}{\nu} = \frac{2\epsilon^{\frac{1}{2}}v_0}{c}\sin\tfrac{1}{2}\theta \tag{5.50}$$

which is obtained from (2.3b), the conservation of energy (fig. 2.23), and (4.70), remembering that $k_l = (2\pi\nu\epsilon^{\frac{1}{2}}/c)$ for a medium of dielectric con-

Brillouin scattering

stant ϵ. Once the acoustic phonon velocities v_0, have been determined the elastic constants of the crystal can be evaluated.

Since there are three acoustic branches (§2.4) having different values of v_0 for a general direction of propagation, three Stokes and three anti-Stokes bands would be expected in the Brillouin spectrum. The Stokes and anti-Stokes bands for a given branch have an almost equal intensity as a result of the small value of the phonon frequency.

Experimental studies of Brillouin spectra have been reviewed by Loudon [5]. Fig. 5.15 shows the Brillouin spectrum of fused quartz for mercury arc excitation. The two Brillouin bands correspond to longitudinal and transverse acoustic phonons. These results on fused quartz have been confirmed and extended by laser studies [570–73].

Stimulated Brillouin scattering studies have recently been carried out by using giant pulse lasers and such studies lead to greatly amplified Brillouin scattering. These studies have been reviewed [574].

5.12 PURE ABSORPTION

It has been seen (§4.4) how the transmission infrared spectrum measures not only absorption but also reflection and emission and sometimes Tyndall scattering effects.

It is possible to measure only absorption by calorimetric measurement [576, 577] but these measurements can lead to errors because of the need to know the specific heat of the crystal [578, 579]. Recently [580] a two laser technique has been developed which allows the calorimetric measurement of absorption for crystals of unknown specific heat.

Plendl has shown that it is possible to evaluate pure absorption directly from vibrational spectroscopic data [360, 581], by using the combined data from the infrared reflection spectrum of the single crystal and the infrared transmission spectrum of an evaporated thin film. Plendl applies (4.51) to the thin film (for which interference effects can be neglected) and substitutes for $R(\nu)$ the reflectivity of the single crystal. Noting that the reflectivity of the single crystal though proportional to $R(\nu)$ makes a smaller contribution to the overall energy balance the variable X is introduced to replace the 1 in (4.51). Thus neglecting emission, (4.51) can be written

$$A(\nu) = X - (R'(\nu) + T(\nu)), \qquad (5.51)$$

where $X > 1$ and $X \to 1$ as the film thickness tends to infinity, and $X \to 1 + R'(\nu)$ as the film thickness tends to zero, and $R'(\nu)$ refers to the

Second order vibrational features

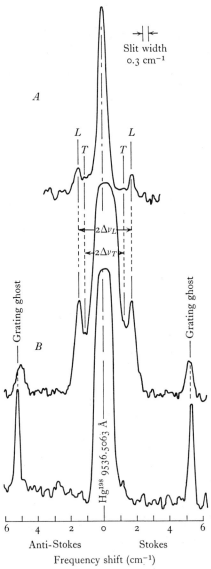

Fig. 5.15. Microphotometer tracing of the Brillouin spectra of two quartz specimens showing both longitudinal (L) and transverse (T) components. Reprinted by permission from Flubacher, Leadbetter, Morrison and Stoicheff [575].

Pure absorption

reflectivity of the single crystal. The maximum value of X may be determined from the maximum value of $(R'(\nu) + T(\nu))$ in a frequency region where $A(\nu) \to 0$. $A(\nu)$ can now be determined from (5.51) and the *specific* or *characteristic absorption* for the single crystal will be $A(\nu)$ divided by the energy entering the single crystal $(1 - R'(\nu))$, and for the thin film will be $A(\nu)$ divided by $(X - R'(\nu))$. Fig. 5.16 shows the characteristic absorption spectrum for thin films of NaCl calculated in this way. It

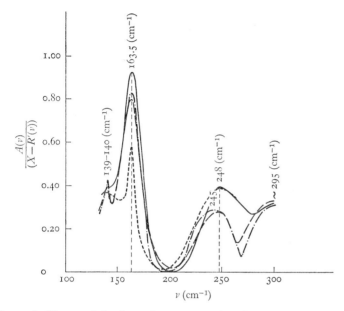

Fig. 5.16. Characteristic absorption spectra of NaCl for various film thicknesses —— = 3.6 μm; .... = 2.3 μm; — — = 1.35 μm; - - - - - = 0.17 μm. (After Prendl, in *Far Infrared Properties of Solids* (ed. Mitra and Nudelman), Plenum Press (1970).)

will be noticed that the longitudinal optical mode is clearly resolved since non-normal incidence is used (§4.10.1). The characteristic absorption spectrum calculated in this way have been shown [360, 581] to give good agreement with the spectrum calculated from elastic constants and the results of neutron scattering experiments.

5.13 ATTENUATED TOTAL REFLECTION

It follows from the discussion of reflection spectra in §4.4 and §4.6 that intense reflection spectra indicate large extinction coefficients (χ). Thus for values of n between 1 and 2 and χ greater than 0.2 reliable optical

Second order vibrational features

constants can be derived from such reflection spectra. The reflection spectra of samples with n and χ outside this useful range can be studied by a new technique developed by Fahrenfort [582].

The new technique involves considering the reflection from the interface between the sample and an optically dense dielectric rather than between the sample and air. If the optically dense dielectric has a refractive index n_2 and the crystal a refractive index n_1, the new refractive index n will be given by

$$n = n_1/n_2. \qquad (5.52)$$

In normal reflection studies (external reflection) $n_2 = 1$ and $n = n_1 > 1$. For perpendicular polarization and $\chi = 0$ (known as S or TE (transverse electric) (§4.10.1)) R rises monotonically with increasing angle of incidence (θ^i) (fig. 4.13) from the value in (4.52) to 100 per cent when $\theta^i = 90°$. For parallel polarization and $\chi = 0$ (known as P or TM (transverse magnetic) (§4.10.1)) R falls to zero at Brewster's angle ($\tan^{-1} n$) and then rises to 100 per cent when $\theta^i = 90°$. In the new technique which involves three media the internal reflection between the sample and optically dense interface is considered. n_2 is normally selected so that $n < 1$. In this case it can be shown that R for both parallel and perpendicular polarization reaches 100 per cent from the value in (4.52) after a small θ^i such that $n \ll \sin \theta^i$ for $\chi = 0$. This total internal reflection effect was first studied by Newton [583]. Newton also noticed that less than total reflection occurred for certain objects placed at the interface. Thus if $\chi \neq 0$ some of the incident radiation will be absorbed by the surface layers of the crystal and the reflection will be less than total. This absorption can be understood if it is noted that sample penetration and beam displacement occurs (fig. 5.17). This penetration occurs because at the interface the incoming wave has an electric field into the sample surface of exponentially decreasing amplitude and the depth of penetration is given [584] by

$$d_p = \frac{cn_2}{2\pi\nu(\sin^2 \theta^i - n^2)^{\frac{1}{2}}} \qquad (5.53)$$

for incident light of frequency ν. The absorbing medium strongly affects the amount of reflection.

It can be shown that the attenuation of reflection is appreciable even for small values of the extinction coefficient. The attenuated total reflection (A.T.R.) spectrum thus represents a transmission spectrum (since the amount of reflection decreases as χ increases). The reflectivity R is related to A' (from (5.23)) by

$$R = 1 - A'd_e, \qquad (5.54)$$

where d_e is the effective thickness. In bulk materials d_p is much smaller than the sample thickness, and d_e is given [584] by

$$d_e = \frac{nE^2 d_p}{2\cos\theta^i}, \quad (5.55)$$

where E is the amplitude to the electric field in the optically dense dielectric. For thin films d_p is much greater than the sample thickness x and d_e is independent of [584] d_p

$$d_e = \frac{nE^2 x}{\cos\theta^i}. \quad (5.56)$$

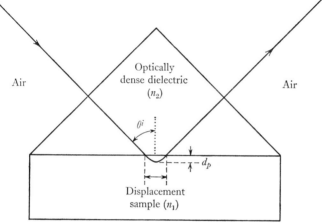

Fig. 5.17. Path of ray of electromagnetic radiation in attenuated total reflection experiment.

For thin films therefore the intensities of bands in the attenuated total reflection spectrum are independent of d_p and ν and are similar to those in transmission measurements. For bulk samples, however, the dependence of the intensity upon d_p and ν causes differences with the transmission measurements. For weak bands the A.T.R. intensities may be divided by the frequency in wave numbers to obtain the approximate transmission spectrum intensity, but for strong bands no simple rule may be applied. For anisotropic substances polarization differences will lead to differences between A.T.R. and transmission spectra. Optical constants may be determined from A.T.R. measurements by using the Kramers–Kronig relation [582, 583]. The whole subject of internal re-

Second order vibrational features

flection has been reviewed [583, 585], including a book by Harrick [584], who also gives a complete discussion of the experimental arrangements.

The optically dense dielectric may be provided by using single crystals such as KRS-5, KRS-6, AgCl, $SrTiO_3$, TiO_2, Ge, Si, and CsI which cover a range of refractive indices (see Harrick [584] for more details). There is often some difficulty in establishing satisfactory optical contact between such dielectrics and the single crystal being studied. It is sometimes helpful in establishing optical contact to slowly crystallize the sample in direct contact with the dielectric. The use of a liquid as the dielectric is possible but is often complicated by liquid absorption.

A.T.R. measurements have been applied to the studies of amorphous solids and powders (Harrick [584] has reviewed these studies). For powders the technique is particularly valuable because the complications caused by scattering can be removed. The powder may be spread or stuck with adhesive tape onto a plate of dielectric. Fig. 5.18 illustrates the spectral improvement that results from using this technique [584, 586].

The theory of A.T.R. for measurements using polarized radiation studies of anisotropic solids has been developed [587]. Recently [588, 589] there have been polarized A.T.R. studies of single crystals. For uniaxial nitrate crystals the ν_3 mode of NO_3^- was found to be radically different for S and P polarization, the P polarization band showing transverse and longitudinal components.

5.14 ABSORPTION BY METALS

It is possible for metals to absorb infrared radiation. For example transverse and longitudinal acoustic phonon bands at about 35 cm^{-1} and 70 cm^{-1} have been observed in the far infrared absorption spectrum of single crystals of lead [590].

If metals are considered to exist in the form of the free-electron gas model then there can be no absorption of infrared radiation since there is no mechanism for the loss of electromagnetic radiation (except under special conditions [591]). It has been shown [592–4] however that even at absolute zero there are two mechanisms that can lead to the loss of electromagnetic radiation.

(i) The electron can acquire an oscillating energy from the electric field in passing through the metal surface, and this energy can be converted into heat by collisions, thus photons are absorbed and phonons created.

Absorption by metals

(ii) The electron can absorb a photon and the resulting electron–hole pair (exciton) can interact with acoustic phonons and hence can generate acoustic phonons. Energy and momentum will be conserved in this simultaneous process. In the high frequency region *all* $K = 0$ acoustic

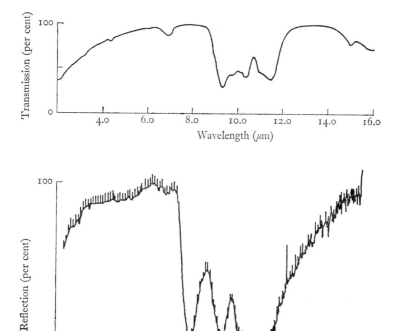

Fig. 5.18. Comparison between transmission and internal reflection powder spectra of augite. Reproduced by permission from Harrick, *Internal reflection spectroscopy*, © John Wiley and Sons, Inc. 1967.

phonons can be generated, but this is not possible at lower frequencies where some of the $K = 0$ acoustic phonon cannot be generated.

The theory of the absorption in lead has been described in detail [595].

Metals can also absorb infrared radiation due to the excitation of waves other than phonons, and this will be discussed in chapter 6.

Second order vibrational features

5.15 THE CHOICE OF SPECTROSCOPIC WINDOWS

Spectroscopic windows are chosen so that they have a high transparency to infrared radiation. It is important that such windows should have no infrared absorption bands in the region of interest. Diatomic cubic ionic crystals are particularly suitable since they have only one absorption and reflection maximum (given in table 4.1) and may be produced commercially as large single crystals. In these compounds the absorption and reflection are appreciable (R is nearly 100 per cent) and thus the cutoff point can be evaluated from table 4.1 remembering that the reflection band stretches from ν_l to ν_t.

Polythene is often used for far infrared studies where nearly all the ionic crystals above have reflection and absorption maxima. Polythene has a medium intensity band at about 80 cm^{-1} but is unsuitable for use above 350 cm^{-1} where it has a number of bands. One might expect the ionic crystals above to be transparent in the very far infrared on the low frequency side of their absorption and reflection maxima, but this is not the case because of appreciable multiphonon absorption. The elimination of multiphonon difference processes by cooling can render the materials transparent at very low frequencies, thus a crystal of CsI has a transmission that goes from zero to 100 per cent below 40 cm^{-1} on cooling from room temperature to 4°K. This phenomenon known as 'supertransparency' together with the details of the multiphonon processes involved has been reviewed by Hadni [597]. Crystalline quartz, silicon and diamond are transparent in the far infrared.

Details which allow the choice of suitable infrared spectroscopic windows are available [596].

5.16 PRESSURE DEPENDENCE OF VIBRATIONAL MODES

When a crystal is subjected to an external pressure the following effects may occur

(i) Strain may develop in the crystal which may change the symmetry of the crystal (for example [598] [598a]) and/or cause mixing between modes [599, 600] and both of these effects may cause inactive modes to become active.

(ii) Frequency shifts nearly always occur. It was seen in §5.2.3 that pseudo-harmonicity caused the fundamental frequency ν' to depend upon

Pressure dependence of vibrational modes

the crystal volume. This pseudo-harmonicity was seen to arise because of the presence of anharmonic terms in the potential energy function and the effect of volume upon the frequency can be given in terms of γ (the damping factor (§4.6) or Grüneisen parameter)

$$\gamma = -\frac{d \log_e \nu'}{d \log_e V}, \tag{5.57}$$

where V is the crystal volume. External pressure will clearly affect this volume and thus ν' and this pressure dependence can be expressed by

$$\left(\frac{\partial \nu'}{\partial P}\right)_T = \gamma \beta \nu', \tag{5.58}$$

where β is the compressibility (§2.8.2). γ may be estimated from Born and Huang theory, assuming a rigid ion model, and evaluation of the volume dependence for a Born–Mayer, or inverse power type potential. Mitra [12] has given the full details.

Experimental studies [601–11] have shown the sensitivity of ν' to pressure, the frequency change being greater for external than for internal modes. For cubic crystals, ν' shifts to higher frequencies on increasing the pressure [601–11], but for non-cubic anisotropic crystals the shift depends upon the particular type of mode [609, 610]. Thus for uniaxial (C_{6v}^4), CdS and ZnO, pressure lifts the degeneracy of the E_2 modes, the low frequency branch going over to the transverse-acoustic branch and *decreasing* in frequency with increasing pressure [608]. This low frequency branch is thus unusual in having a negative γ value. Negative γ values may arise because of an anomalous negative volume coefficient of thermal expansion, and/or a decrease in stiffness of the lattice to a particular mode under compression, preceding a first order transition [612].

The pressure dependence of acoustic phonon frequencies have been measured by ultrasonic methods [612], neutron scattering [613–15], and tunnelling spectroscopy [616].

(iii) Changes may occur in the intensities of bands due to the pressure induced dipole moment and polarizability changes. This effect can be useful to enhance the intensity of weak defect modes (§5.19) and for distinguishing between asymmetric and symmetric vibrations since pressure affects the intensity of one mode more than the other [611].

(iv) Strain may disorder the crystal and, if this happened, $K \neq 0$ fundamentals could become active (§5.18).

Second order vibrational features

Bradley has reviewed the use of high pressure methods [617], and far infrared studies at high pressures [618]. High pressures have been found useful in speeding the growth of single crystals [619].

5.17 THE EFFECT OF APPLIED FIELDS

5.17.1 Induced infrared absorption and Raman scattering

An externally applied electric field can give rise to induced infrared absorption and Raman scattering. The electric field will induce a dipole moment in the sample given by (3.71) where E now applies to the externally applied static field. An induced effective ionic charge $\partial X/\partial u$ (see 4.34) therefore arises and since the intensity of the induced infrared absorption with be proportional to the square of this charge then

$$I_{\text{induced}} = \text{Constant} \left| \left(\frac{\partial \alpha_\xi}{\partial Q_k} \right) E \right|^2. \qquad (5.59)$$

Thus the application of an electric field produces a first order phonon spectrum in diamond [620].

The externally applied electric field can remove the inversion symmetry of a crystal and thus allow first order Raman scattering, an effect observed in $KTaO_3$ [55, 621, 622] and $SrTiO_3$ [55, 621, 622]. Mitra [12] has discussed these effects in more detail.

The externally applied electric field also causes a change in the mode frequencies via the anharmonic forces. The shift may be proportional to the electric field (in non-centrosymmetric systems) [623] or proportional to the square of the electric field (cubic systems). The shifts are greatest in ferroelectric crystals (where there is the greatest amount of anharmonicity).

5.17.2 Enhancement of Raman intensities in a magnetic field

Electronic energy levels are susceptible to magnetic fields. The energy levels in the magnetic field are often split into a number of closely spaced levels known as Landau levels. When the energy difference between two Landau levels is equal to a longitudinal phonon energy, mixing between the states occurs due to electron–longitudinal phonon coupling. This coupling has been observed to give rise to enhanced Raman scattering by longitudinal optical phonons in CdS [624]. The theory has been discussed [624a].

The effect of applied fields

5.17.3 Non-linear Raman effects in large electric fields

When the monochromatic radiation used in Raman studies is produced by a giant pulse laser the sample is subjected to very large electric fields. Stimulated Raman scattering and inverse Raman scattering can be observed under such conditions (see the review by Long [625]), as well as multiphoton Raman scattering.

(3.71) should appear as a power series in $|\mathbf{E}|$

$$|\mathbf{X}| = \alpha_\xi \cdot |\mathbf{E}| + \tfrac{1}{2}\beta_\xi|\mathbf{E}|^2 + \tfrac{1}{6}\gamma_\xi|\mathbf{E}|^3 + \ldots \quad (5.60)$$

Normally the second and higher terms are not important and thus (3.71) is used for the Raman effect. In very large electric fields however these higher terms must be considered. The second term which depends upon the first hyperpolarizability tensor β_ξ, gives rise to three photon Raman scattering (the hyper Raman effect). Two incident photons of frequency ν_1 and ν_2 are considered (usually $\nu_1 = \nu_2$) thus (see fig. 2.23 and replace ν by $\nu_1 \pm \nu_2$):

$$\nu_f = \nu_1 \pm \nu_2 \pm \nu'. \quad (5.61)$$

Thus ν_f may equal 0 (Rayleigh scattering ($\nu' = 0$) with $\nu_f = \nu_1 - \nu_2$ and $\nu_1 = \nu_2$), $\nu_1 + \nu_2$ (Rayleigh scattering), $\nu_1 + \nu_2 \pm \nu'$ (hyper Raman Stokes and anti-Stokes terms), and ν' (Raman scattering with $\nu_f = \nu_1 - \nu_2 \pm \nu'$ and $\nu_1 = \nu_2$). This hyper Raman effect may be found discussed in more detail elsewhere [625, 626]. Selection rules, which will now depend upon the irreducible representation of β_ξ instead of α_ξ, have been evaluated for most point groups [627, 628]. An interesting feature of these selection rules is that all infrared active bands are also hyper Raman active.

The third term of (5.60) which depends upon the second hyperpolarizability tensor γ_ξ gives rise to four-photon Raman scattering. Three incident photons of frequency ν_1, ν_2, and ν_3 are considered (usually $\nu_1 = \nu_2 = \nu_3$) thus (see fig. 2.23 and replace ν by $\nu_1 \pm \nu_2 \pm \nu_3$)

$$\nu_f = \nu_1 \pm \nu_2 \pm \nu_3 \pm \nu'. \quad (5.62)$$

Selection rules, which depend upon the irreducible representation of γ_ξ, have been evaluated for most point groups [628].

5.18 HYDROGEN BONDING

In crystals if hydrogen bonding, which may be represented as X—H...Y occurs, the X—H internal stretching mode is often very broad. Thus the —O—H stretching vibration in such crystals increases in breadth as

Second order vibrational features

the strength of the hydrogen bonding increases, and the band often shows fine structure. The width of these bands may be explained by the factors discussed in §5.4. Correlation field splitting in hydrogen bonded systems is generally very large, as might be expected because of the strong interaction between internal modes through hydrogen bonding. Thus the correlation field splitting in H_2O_2 is 93 cm^{-1} [4] and so unresolved correlation field components will lead to appreciable band width. Dilute isotopic solid solution studies show that the width of the —O—H stretching vibration in HOD is only 80 cm^{-1} (§5.5.2 has described this technique for elimination the effect of correlation field splitting). In hydrogen bonded systems a further factor has been considered to contribute to the band width, namely the width associated with the double potential minimum. This double potential minimum is considered to arise if a proton transfer model is considered. Thus there will be one potential minimum for the proton near X, and another with the proton near Y. Doubling of the —X—H internal vibrational mode has been considered as evidence for 'proton tunnelling' through the potential barrier between the two potential wells, though such doubling could be due to correlation field splitting, multiphonon processes, etc., and in any case such doubling does not necessarily arise for the double-well potential model [4, 629]. The potential barrier can be considered in terms of its effect on the lifetime of the hydrogen atom in the well, and thus contribute to the width of the mode, such an effect having been used to explain the width of bands in the ice spectrum [4, 630]. There have been many studies of the vibrational spectra of hydrogen bonded crystals based upon the double-well proton-transfer model and these have been fully reviewed elsewhere [629, 631–3]. Other models include a double-well proton-transfer model, with interaction of the X—H and X–Y modes [634, 635], and a similar model with strong interaction with external modes considered to infinite order by construction of a quasi-particle analogous to the small polaron (chapter 6) [629].

Relationships have been derived between the X–H internal stretching mode frequency and the H...Y distances by using potential functions for the van der Waals' forces involved [636].

A number of general reviews of the subject are available [629, 631–3, 637, 638].

Crystal disorder and defects

5.19 VIBRATIONAL SPECTROSCOPIC EFFECT OF CRYSTAL DISORDER AND DEFECTS

The effect of crystal disorder and defects upon the lattice dynamics was discussed in §2.7.4. It was seen that these effects destroy the translational symmetry of the crystal and break down the conservation of wave vector rule. This means that $K \neq 0$ phonons can give rise to one phonon fundamentals and unit cell group analysis cannot therefore be used to predict activities in the infrared and Raman spectrum.

5.19.1 *The spectra of disordered crystals*

There are many crystals that exhibit disorder, and often this disorder takes the form of molecules on, or nearly on, regular lattice positions but oriented irregularly to the lattice. Such crystals are intermediate between regular crystals and glasses which do not have the positional order of disordered crystals. The breakdown of conservation of wave vector in disordered crystals and glasses causes *all* phonons [639] whether optical, acoustic, transverse, or longitudinal to become infrared and Raman active. In glasses the situation is more complicated than in disordered crystals because the lack of positional order leads to irregularity in the normal vibrations. The number of observed vibrations will be limited in practice by the number of critical points in the one phonon density of states. The intensity of the observed bands will depend upon the vector sum of the $\partial \mu_g / \partial Q_k$ values for the individual molecules or groups, and because of their irregular orientation the intensity will depend upon the particular sample disorder.

The far infrared spectra of ice (I_c) [640] and amorphous As_2Se_3 [641] provide examples of the spectra of disordered crystals. Ice has a space group O_h^7 with only two molecules in the unit cell. Only one degenerate external vibration would therefore be expected and unit cell group analysis predicts that this mode would be Raman active and infrared inactive. Crystal disorder causes the mode to become infrared active, and the spectrum shows fundamentals for a critical point at the zone boundary with the transverse optical mode at 229 cm^{-1}, longitudinal optical mode at 190 cm^{-1}, longitudinal acoustic mode at 164 cm^{-1} and transverse acoustic mode at 65 cm^{-1}.

Random polymers [201, 642] also exhibit the features of disordered crystals. There have been many studies of the vibrational spectra of glasses [643-8], which because of the lack or orientational and positional order behave more like gases.

Second order vibrational features

5.19.2 Types of crystal defect

Crystals can exhibit a number of point defects, and a complete discussion of such defects can be found in most textbooks of solid state physics [24, 649]. Table 4.5 lists many of these defects. Most of these defects involve electrons and are known as *colour centres* because they often absorb visible light. Such transitions are outside the scope of this book and have been fully reviewed elsewhere [649]. These defects also cause effects in the vibrational spectroscopic region due to localized, non-localized and resonant modes and these features will be discussed below.

TABLE 5.5 *Point defects in solids*

Name of defect	Cause
Substitutional defect	One atom, or ion replaces another
Schottky defect	Lattice vacancy, with an atom or ion going to the surface
Frenkel defect	Lattice vacancy, with an atom or ion going into an interstitial position in the lattice
F centre	Lattice vacancy caused by the removal of a negative ion, and this vacancy replaced by an electron
F_A centre	An F centre, but with one of the positive ions surrounding the F centre replaced by an ion of the same charge but different mass
M centre (F_2 centre)	Two adjacent F centres
R centre (F_3 centre)	Three adjacent F centres
N centre (F_4 centre)	Four adjacent F centres
F_2', F_3', F_4' centres	Electrons trapped by M, R, and N centres respectively
F_2^+, F_3^+, F_4^+ centres	Holes trapped by M, R, and N centres respectively
Anti-F centre	Hole trapped at a positive ion vacancy
V_K centre	Hole trapped by a pair of negative ions (preferred to an anti-F centre)
V_{KA}, $V_{KA'}$, V_F, V_H centres	Modified V_K centres
H centre	Doubly ionized molecule ion occupies the site of its singly charged atomic ion
U centre	H^- ion located at a negative ion vacancy in an alkali halide crystal.
H_i^- centre	H^- ion located at an interstitial site in the vicinity of an anion vacancy
H_i^0 centre	Hole trapped by H_i^0 centre

5.19.3 Substitutional defects in elemental crystals

Elemental crystals will not, of course, give rise to first order infrared and Raman effects because only acoustic phonons are possible. The introduction of heavier or lighter substitutional defects can cause non-localized or host modes that have the effect of allowing the first order elemental crystal spectra to be studied [464]. This first order spectrum can also be studied by introducing defects by electron or neutron irradiation [168] (e.g. in studies of diamond, silicon and germanium), but since the nature of the defect is then not known in detail it is often preferable to introduce a specific substitutional defect.

In general, for any type of crystal, substitutional defects will allow all phonons to be infrared and Raman active, but their infrared intensity depends upon the dipole moment change associated with the defect, $\partial \mu_\xi^d / \partial Q_k$, and the density of states $S(\nu)$ at the non-localized or host mode frequency ν. Thus the intensity of the non-localized host mode will be given by a modified equation (5.36) [650]

$$A_\xi'' = \frac{1}{n} \left| \frac{E_l}{E} \right| \frac{A^d \pi}{c^2} \sum_k \left(\frac{\partial \mu_\xi^d}{\partial Q_k} \right)^2 S(\nu), \qquad (5.63)$$

where A^d now refers to the number of defects per unit volume. Similar considerations apply to the Raman spectra [155].

For elemental crystals $\partial \mu_\xi^d / \partial Q_k$ is positive because there is a charge associated with the defects, though it is not strictly necessary that the defect atom should actually add charge as long as a local redistribution (known as overlap distortion) of charge occurs to make $\partial \mu_\xi^d / \partial Q_k$ finite. The calculation of $\partial \mu_\xi^d / \partial Q_k$ for various types of defects has been reviewed [168, 3], and the theory of inert gas crystals has been discussed [168, 651].

The defect-induced spectra of silicon provides a good example [155, 168, 652–4]. A boron defect (lighter than silicon) causes silicon modes together with a localized mode. A phosphorus defect (heavier than silicon) causes a non-localized mode. The defect-induced spectra of inert gas crystals by neutral substitutional defects arises through overlap distortion. The one-phonon spectrum of argon was first studied using a xenon defect [655], and the spectrum found to agree well with the calculated density of states for argon (since a range of phonons with different K values are excited). Later studies have involved the one-

Second order vibrational features

phonon spectrum of argon and krypton with monatomic (Ar, Kr, Xe) [656], diatomic (CO, N_2, D_2, H_2) [656, 657], and polyatomic (CH_4) [656] impurities. It has sometimes been found necessary to use a high pressure technique to obtain a suitable matrix [656]. Polyatomic impurities induce bands of higher frequency than those of monatomic impurities presumably because combination bands between induced phonons of the host and librons and phonons of the polyatomic impurity are observed. Localized modes are observed with suitable impurities (e.g. for D_2 [656, 657]).

The spectroscopic properties of defects in the semiconductor GaSb, InSb and GaAs have been fully reviewed [168].

5.19.4 *Matrix isolation – a special case of substitutional defects in elemental and molecular crystals*

Matrix isolation studies have become increasingly important [658] not only as a means of studying short-lived species by trapping them in an inert solid, but also as a means to investigate the spectrum of a material without encountering the usual complicating features of solid state spectra. The spectrum of a matrix isolated material more closely resembles that of a gas than of a solid since the trapped molecule is entirely surrounded by inert solid and therefore only the term $\sum_k V_k$ in the potential energy expression (1.1) is involved.

In addition the bands due to the trapped molecule will be much sharper than in solid state studies because of the lack of coupling with phonons, virtual width, unresolved solid state splittings, and a superimposed reflection maximum. The trapped molecule can often exhibit further similarities with a gas by rotating in its site and exhibiting this rotation as fine structure on the vibrational bands. This rotation is affected by the surroundings however and the theory for such rotation has been fully discussed [659–61]. Rotation–translation coupling [659] can give rise to a localized mode of the external type discussed so far.

If the matrix is capable of any one-phonon transitions, coupling between the localized molecular mode of the matrix isolated material and the external modes of the matrix must be considered, i.e. the term V_{Ek} in the potential energy expression (1.1) must be included. Since substitutional defects always allow first order elemental or molecular crystal spectra such coupling, which may lead to band shifting, band splitting (§5.5.3), or combination bands, must *always* be considered. In the case of a matrix

Crystal disorder and defects

of some molecular crystals (such as CO_2) the possibility of coupling with internal modes (which may be very much greater than that between external modes because of a smaller frequency difference) must be seriously considered, together with the greater number of combination band possibilities.

Combination bands between the localized molecular mode of the matrix isolated material and $K \neq 0$ acoustic modes of the matrix have been experimentally observed [662, 663], though it must be remembered that such combinations require the potential energy expression to contain cross-terms between matrix isolated and matrix molecules, and thus the combination band possibilities are less than in a homogeneous solid.

The possibility that strain may develop in the matrix isolated species (with the possible effects discussed in §5.16) as a result of the packing in the matrix should also be considered.

5.19.5 Substitutional defects in ionic crystals

For ionic crystals it is no longer necessary for a charge or overlap distortion to be associated with the defect in order that $\partial \mu_\xi^d / \partial Q_k$ be positive. This is because the pure crystal is already charged (and of course already shows first order infrared and Raman effects) and thus defect-induced perturbations of the motions of the normal modes can lead to a positive $\partial \mu_\xi^d / \partial Q_k$ and thus additional first order infrared and Raman processes. Unlike the elemental and molecular cases discussed above defect induced modes in ionic crystals are liable to be intense, because $\partial \mu_\xi^d / \partial Q_k$ will almost certainly be large, since it involves a perturbation of the motions of charged particles.

Localized, non-localized, and resonant modes involving a variety of substitutional defects in ionic crystal have been studied.

The U centre is a colour centre giving rise to an ultraviolet absorption band. Infrared absorption occurs due to the localized modes of the very light H^- or D^- ions, which give the expected $\sqrt{2}$ frequency ratio for this localized mode. The extensive experimental work on U centre infrared absorption has been fully reviewed [155, 168, 664]. The localized mode frequency depends upon the host lattice varying from 363 cm^{-1} in CsBr to 1025 cm^{-1} in LiF (for H^-). Combination bands (both difference and summation) have been observed involving the localized U centre mode and host phonons with an intensity dependent upon the single phonon density of states for the host lattice [664, 665, 473, 155]. Ultraviolet irradiation destroys U centres and creates a new kind of centre called

Second order vibrational features

an H_i^- centre where an interstitial H^- ion occupies different interstitial sites in the vicinity of an anion vacancy giving a different localized mode frequency for each site [666]. Gap modes are observed for U centres in lithium hydride. Raman scattering from U centres is particularly striking since the alkali halide host crystals are Raman inactive. The substitutional defect, such as the U centre, induces Raman scattering because the breakdown of translational symmetry leads to the loss of a centre of inversion for the ions adjoining the defect. The Raman scattering from defects has the advantage of being able to determine the symmetry properties of the induced host mode. The theory of defect Raman scattering has been well developed [155, 667, 668].

Other defects also give rise to induced infrared absorption and Raman scattering. Examples of infrared studies include the Ag^+ defect [664] in KCl, KBr, KI, and NaCl, and the Li^+, Cu^+, Ag^+, and Eu^{2+} defects in KBr, NaCl, KI, and MnF_2 respectively [669, 670], where resonance modes are observed in far infrared studies. The temperature dependence of such resonance modes lead to useful information about the appreciable anharmonic nature of the impurity binding forces [155, 669] (in the case of Li^+ in KBr for example $\gamma = 50$!). The appreciable anharmonicity and low force constant characterizing a light ion resonant mode (such as Li^+ in KBr) is a result of the defect-induced perturbations of the motions of the normal modes discussed earlier which causes the defect ion to go off-centre. For a small ion the crystal energy is reduced by polarizing the lattice ($\partial \mu_\xi^d / \partial Q_k$ positive) by off-centre movement without an appreciable increase in the repulsive energy with the host [155, 670]. The off-centre behaviour leads to a double potential minimum (§5.18) which can give rise to tunnelling transitions between the vibrational energy wells and the tunnelling levels are split in applied electric fields [155, 67]. Such tunnelling transitions occur in the very far infrared. Molecular impurities such as OH^- [671], NO_2^-, CO_3^{2-}, NO_3^-, CO_2^-, O_2^-, BO_2^-, BH_4^-, and NH_4^+ [672–6] in alkali halides exhibit the same properties as off-centre ions but their spectra are more complicated. Localized gap modes in such systems have been reviewed [168].

Other infrared examples include the studies of localized and localized gap modes of NCO^- and CN^- in alkali halide lattices by observing the combination bands between these localized external modes and the localized molecular modes (the internal modes) of NCO^- and CN^- [13]. Examples of Raman studies include the induced host mode near 200 cm^{-1} in KCl produced by doping with I^-, Br^-, Li^+, Na^+, Cs^+, Rb^+, and Tl^+

Crystal disorder and defects

impurities [677-9]. Localized modes due to O_2 [680, 681], S_2^-, S_3^-, Se_2^-, SeS^- and N_2^- in alkali halides [682] have been observed.

There have been a number of studies of the vibrational spectra of mixed crystals of the type $AB_{1-x}C_x$. Two types of behaviour have been observed in such diatomic systems. In the first type of behaviour one mode is observed of fairly constant intensity with a frequency that varies continuously from one end member to the other. In the second type of behaviour two modes are observed whose intensity is proportional to mole fraction of the relevant component and whose frequencies are close to those of the end members. The two mode behaviour can be explained as being caused by localized and gap modes where substitution of C for B in AB produces a gap mode and substitution of B for C in AC produces a localized mode (for $mass_C > mass_B$). The one mode behaviour occurs because the conditions for the localized and gap modes are not fullfilled. Chang and Mitra [683] have predicted that $mass_B < \mu'_{AC}$ (where $\mu'_{AC} = mass_A \, mass_C/(mass_A + mass_C)$) if two-mode behaviour is to be observed, though this rule only applies strictly to a one-dimensional chain and not a three-dimensional crystal [684]. Mitra [12, 685] has reviewed examples and theoretical approaches to these mixed crystal systems.

F centres [686] can be considered as substitutional defects of vanishing defect mass, though they behave as if they had a finite mass because of the existence of long range forces in the lattice. They thus give rise to induced infrared absorption and Raman scattering. Infrared studies have involved the observation of gap [668, 687] and strongly resonant modes [668] in KBr and KI. The gap mode can be explained by a molecular model [688]. Raman studies have involved the observation of induced host modes and gap modes in alkali halides [689, 690].

Bound polarons (an electron attracted to a positive ion site, to be discussed in §6.2.3) can be considered as substitutional defects, the lattice polarization that results leading to a change in the local force constant and thus a localized mode. An induced localized mode in CdF_2 has been observed by Raman scattering [691].

Raman scattering can arise from localized mode polaritons, in the same way as it could arise in pure crystals (§4.8.2.) if such localized modes have a transverse infrared active character. The theory for such localized mode polaritons has recently been discussed [692].

Localized modes may be split due to electron-phonon interaction between the localized phonon and the electron of the ultraviolet or visible process [692a].

Second order vibrational features

Phonons can be elastically scattered by defects, boundaries dislocations (which cause appreciable scattering) etc., with a scattering cross-section that varies as the fourth power of the phonon frequency. Such phonon scattering has an important effect upon the thermal conductivity. Phonon scattering in various crystals caused by various defects has been reviewed [693, 694, 695].

6 Spectroscopic effects other than vibrational transitions in the vibrational energy region

6.1 INTRODUCTION

In addition to phonons there are a number of elementary excitation waves that can occur in solids, and many of these waves can be excited by electromagnetic radiation. Usually the electromagnetic radiation that can excite such waves lies outside the vibrational energy region, and thus a full discussion of these waves is outside the scope of this book. In this chapter the excitation of these waves by electromagnetic radiation in the vibrational energy region is discussed.

6.2 EXCITONS AND ELECTRONIC TRANSITIONS

When a collection of atoms or molecules come together to form a solid a new ground electronic state for the solid arises which has a lower energy than the ground state of the individual atoms or molecules. The excited electronic states for the solid will consist of a number of bands which replace the electronic energy levels of the individual molecules. The band containing filled orbitals is called the valence band. Two types of band containing empty excited orbitals are possible depending upon the energy of the excited orbitals and the nature of the solid. In one type of band the excited states are lower lying and are concentrated on one atomic or molecular unit in the solid. In the other type of band the excited states have a higher energy corresponding to electron ionization and such a band is known as the conduction band, since the electron is free to travel throughout the entire crystal, and hence to conduct electricity. When an electron is excited from the valence band to the excited state there is an apparent positive charge, or hole, left in the valence band. The electron and the hole are attracted by Coulomb forces, lowering the energy of the system, and leading to new energy levels beneath the excited state bands. The bound electron–hole pair is known as an *exciton* and the new energy levels exciton levels. Tightly bound excitons are associated with the lower lying energy levels that are concentrated on one atomic or molecular unit

Other effects in vibrational energy region

in the solid. Weakly bound excitons are associated with the conduction band and thus the electron–hole interparticle distance is large in comparison with a lattice constant. In both cases the excitation spreads outwards through the crystal like a wave. Thus in the same way as lattice vibrations were considered in the solid as propagating phonon waves, electronic transitions are also considered in the solid as propagating exciton waves. Like phonons, excitons have an energy that is dependent upon the wave vector, and obey the conservation of energy and wave vector. They are termed S, P, D, etc., referring to the appearance of the relevant exciton band (i.e. the path that the electron takes around the hole). The exciton, being a neutral particle, cannot carry current, though it can perturb electrical conductivity (just as phonons can perturb thermal conductivity). Excitons are readily polarized by an applied field leading to an energy dependence upon the applied field (D excitons are more strongly polarized than P excitons, which are more strongly polarized than S excitons). Excitons may be excited (as long as (3.69) is finite, where p_ξ now refers to r_ξ see (3.68)) by a direct process ($k_{\text{photon}} = k_{\text{exciton}} \approx 0$), or an indirect process involving the emission of a phonon

$$(k_{\text{photon}} = k_{\text{exciton}} + K_{\text{phonon}}),$$

and thus excitons with a range of wave vectors can be excited in first order processes. The exciton energies are normally in the visible or ultraviolet regions of the spectrum. Dexter and Knox [696] have given an excellent introduction to the subject, and the subject is covered in detail in a number of books and reviews [697–704].

6.2.1 Transitions in paramagnetic species

In §5.19.2 it was seen how most defects give rise to transitions in the visible and ultraviolet spectral regions. The electronic states of an ionic substitutional defect will be affected by the interaction with the host crystal and this crystal field may remove (i.e. site group effect) the degeneracy of some of the free ion electronic states. The same crystal field effect can occur in the pure crystal. In paramagnetic ions (first transition metal series and the rare earths in particular) the lifting of electronic state degeneracies often leads to energy separations in the far infrared.

In infrared studies, transitions between these levels (which are usually d to d or f to f) are usually Laporte forbidden but can occur because the crystal field causes electronic state mixing. Examples include far infrared

Excitons and electronic transitions

studies of transitions due to Ti^{3+} and V^{4+} in Al_2O_3 [705], Fe^{2+} in ZnS [706], and Cr^{3+} in beryl (synthetic emerald) [707], and a number of transitions in rare earth ions (reviewed by Hadni [708]). In Raman studies transitions between these levels are allowed and have also been observed in particular in many studies of trivalent rare earth ions in garnet crystals. This electronic Raman effect is governed by different selection rules than those that apply to the vibrational Raman effect, because the scattering tensor may be antisymmetric rather than symmetric. For an antisymmetric tensor the equalities in (3.76) no longer hold and thus all nine tensors are required. Linear combinations of these tensor components can be made that under an arbitrary rotation around a particular axis transform into themselves, and such combinations are called irreducible tensor components. Koningstein [709] has given a table of such tensors and a list of their transformation properties for most point groups (which refer to the site group in the crystal field). There is a close analogy between the three sets of irreducible tensor components and the s, p, and d atomic orbitals. In evaluating the selection rules for a particular site care must be taken to ensure that the irreducible tensor contains the angular momentum of one of the states involved in the electronic transition. Electronic Raman scattering can be as large or larger than the vibrational Raman scattering, and unlike the vibrational Raman effect can have degrees of depolarization (ρ'') anywhere in the range 0 to infinity. Koningstein [709, 710–12] has fully described the theory of the effect and has reviewed the experimental work, and Chiu has discussed the theory of a novel electronic Raman effect where the incident and scattered photons are of different parity [713].

The electronic nature of these transitions can be clearly demonstrated by examining the effect of a magnetic field upon the spectra. The Zeeman effect [706, 708, 709] causes splitting and shifting of the electronic energy levels. Thus for transitions where the ground state decreases in energy with increasing magnetic field the intensity of the transition increases and vice versa [709].

6.2.2 Transitions in superconducting species

For a normal conductor the valence electrons are the conduction electrons, and there are no energy gaps for transitions to occur. The superconducting state is an ordered state where the conduction electrons are in the form of closely associated pairs. This state, characterized by zero resistance occurs when the conductor is cooled to low temperatures (the

Other effects in vibrational energy region

transition temperature is usually of the order 0.1 to 10 °K). There is an energy gap (of the order 10–30 cm^{-1}) between the electrons in the superconducting state and electrons in the normal conducting state caused by the breaking of the closely associated electron pairs. The magnitude of the gap depends upon the temperature and as more electrons are excited across the gap by increasing temperature the gap falls to zero energy at the superconducting transition temperature T_s. The magnitude of the gap at absolute zero is about $3.5kT_s$. Absorption due to transitions across this energy gap has been observed in far infrared studies [714, 55] and has been fully reviewed by Tinkham [774]. Raman scattering from the superconductor surface due to inelastic scattering from the gap should also be observed, and the theory for such scattering has been discussed [715].

6.2.3 Polarons

Another type of electronic transition can occur if an electron is *added* to a crystal lattice. The added electron interacts electrostatically with the ions, atoms or molecules in the host lattice leading to a local longitudinal polarization field. This deforms the lattice creating a 'polarization cloud' that is carried by the electron as it travels through the lattice and consists of an electron and a cloud of longitudinal phonons, the quantum of energy in such a wave being known as a *polaron*. Another type of polaron, known as an electronic polaron, occurs when the cloud of phonons is replaced by a cloud of excitons [716].

The cloud of phonons that accompany the electron results in an effective increase in the mass of the electron. The interaction between the electron and the lattice is expressed in terms of a coupling constant α given by

$$\alpha = \frac{2(\text{deformation energy})}{h\nu_l}, \qquad (6.1)$$

where ν_l is the longitudinal optical phonon frequency for the phonons of the phonon cloud at $K \approx 0$, thus $\frac{1}{2}\alpha$ represents the number of phonons that surround a slow-moving electron in a crystal. As expected α has a much greater value for ionic compounds than covalent compounds (e.g. 5.6 for KCl and 0.014 for InSb). The value of α will clearly determine the extension of the polarization cloud induced by the added electron. When α is large ($\alpha > 10$), the polarization cloud would be small and of the order of the lattice constant, because the effective mass of the polaron will be so large that the kinetic energy of the polaron is too small to escape the energy well about an atomic or molecular centre. Such a polaron is known

Excitons and electronic transitions

as a *small polaron*. On the other hand when α is small ($\alpha < 2$), the polarization cloud will be large and extend over a number of lattice sites. Such a polaron is known as a *large polaron*. In pure crystals the polaron is known as *free*, but in crystals with defects or disorder the polaron may be bound to the defect or disorder and such polarons are known as *bound polarons*. Polarons, like excitons, are termed S, P, D, etc., according to the path that the electron takes (i.e. the angular variation of the electron probability density).

A full discussion of polaron theory is beyond the scope of the book [717, 719]. The theory of the spectroscopic properties of polarons has been reviewed [717, 720]. Various types of transition are possible depending upon the type of polaron. In the case of small free polarons the polaron can hop from one molecule or ion to another, the need to satisfy the Franck–Condon principle (which means that the lattice polarization is unaffected) leading to an energy requirement of four times the activation energy. This hopping motion will transport electrical charge and the conductivity reaches a maximum at a frequency corresponding to the spectroscopic transition. Small free polaron transitions have been observed in the near infrared in the doped (to render semiconducting) crystals $BaTiO_3$ [721, 722], $LaCoO_3$ [723, 724], $SrTiO_3$ [725], TiO_2 [726, 727], NiO and CoO [728]. Since the polaron consists of an electron and a cloud of longitudinal phonons, polaron transitions are accompanied by the emission and absorption of optical phonons. At high temperatures many phonons participate in the hopping transitions and little phonon structure is observed on the polaron band. As the temperature is lowered, multiphonon processes diminish and phonon structure becomes more pronounced because only a few phonons will participate in the hopping transitions. The relaxation time that determines the width of the polaron transition depends upon the scattering of polarons by phonons. At low temperatures only acoustic phonons have the low energy required and the width becomes less.

Spectroscopic transitions that involve the transfer of an electron from a lower to a higher oxidation state of the same or a different element are known as intervalence transfers. Such transitions can be considered as small polaron transitions though they will not be discussed further here since such transitions usually occur in the visible region. They have been fully reviewed by Hush [729].

In the case of small bound polarons the polaron can hop about the defect, and transitions in such systems have been observed in the far

Other effects in vibrational energy region

infrared. Infrared studies [730] of CdF_2 doped with rare earth ions show such a transition at 180 cm^{-1}, together with a localized phonon mode of the lattice induced by the polaron that can be observed in Raman studies [731] (§5.19).

Large free polarons can also give rise to transitions [720] of the Franck–Condon type involving transition from S to P polarons, and also transitions involving states where there is phonon scattering of the polaron. Transitions that do not obey Franck–Condon rules and involve lattice relaxation have been discussed [720]. Transitions due to large free polarons have been observed in CdO [732] in the near infrared.

Large bound polarons give rise to transitions in the far infrared region. When silver bromide is radiated with ultraviolet radiation an electron is excited into the conduction band and becomes trapped at some positive centre. Absorption due to an S to P polaron transition occurs at 168 cm^{-1}, which is subject to Zeeman splitting, and transitions due to combinations of this transition with an optical phonon and transitions due to an ionized polaron are also observed [733, 734].

6.3 PLASMONS

Electrons in the conduction band of a crystal are no longer deflected by the atoms or molecules of the crystal, and are free to move as an electron gas. Normally the crystal is neutral but if this electron gas is displaced uniformly upwards away from the positive core by an amount η, opposite charges of value $\eta n e$ (where n is the number of electrons of charge e) develop on either side of the surface between the electron gas and the core leading to an electric field $\mathbf{E}_p$ of $4\pi\eta n e$ which acts as a restoring force. This longitudinal excitation of the electron gas acts as a simple harmonic oscillator of frequency ν_p known as the *plasma frequency*, which can be calculated from the equation of motion

$$nm^*\frac{d^2\eta}{dt^2} = -ne|\mathbf{E}_p| = -4\pi\eta n^2 e^2, \qquad (6.2)$$

where m^* is the mass of the conduction band electrons, so assuming simple harmonic motion

$$\frac{d^2\eta}{dt^2} + 4\pi^2\nu_p^2\eta = 0 \qquad (6.3)$$

$$\therefore \quad \nu_p = \left(\frac{ne^2}{\pi\epsilon_\infty m^*}\right)^{\frac{1}{2}}, \qquad (6.4)$$

where the dielectric constant of the medium has been added. The energy

Plasmons

in this excitation wave is quantized as a unit called a *plasmon*. The plasmon just described is known as a *longitudinal plasmon*. Another type of plasmon known as the *transverse plasmon* or dressed photon which consists of a photon surrounded by an electron cloud is possible. The transverse plasmon is a transverse oscillation of an electromagnetic wave modified by fields arising from the collective electron response [735], energy being transferred to electrons moving perpendicular to the wave.

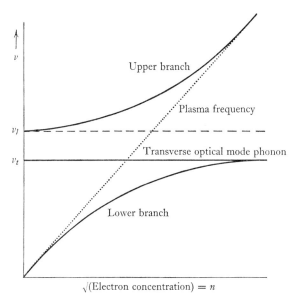

Fig. 6.1. Variation of the frequency of the $K = 0$ coupled longitudinal plasmon–longitudinal optical phonon with the electron concentration.

The Raman scattering of incident radiation by plasmons has been treated theoretically [736, 737]. Surface plasmons [738] can couple with photons and this can be observed in reflection experiments [739–41]. Plasmons have been determined experimentally in Raman scattering from coupled plasmons involving the conduction electrons of various semiconductors. The coupled longitudinal plasmon longitudinal optical phonon arises in suitably doped GaAs when the plasma frequency equals the phonon frequency, in a way rather analogous to the formation of the polariton from phonon–photon interaction. Fig. 6.1 illustrates how the coupled phonon—plasmon exhibits two branches with frequencies that depend upon the electron concentration n. The coupled phonon–plasmon

Other effects in vibrational energy region

can be observed in infrared reflection experiments [742], and in Raman experiments [743, 744] where scattering from the upper and lower branches can be observed by suitably varying n. The coupled transverse plasmon–transverse phonon can also arise in doped GaAs, such an excitation being known as a *plasmariton* [745] because of the analogy of this with a polariton. Like the polariton, the plasmariton can only be observed in near forward scattering studies (§4.8.2) [746].

6.4 MAGNONS

Paramagnetic substances can sometimes under suitable conditions exhibit magnetic behaviour by aligning the spin of their unpaired electron(s) in a ferromagnetic, ferrimagnetic, or antiferromagnetic way. In the ferromagnetic arrangement the spin of the unpaired electron(s) of adjacent paramagnetic atoms are aligned parallel throughout the crystal (fig. 6.2(a)). In the ferrimagnetic arrangement two or more types of paramagnetic atoms are present which can have the spins of their unpaired electron(s) aligned parallel or antiparallel to one another but there is a net magnetic moment for the whole crystal as there was in the ferromagnetic case

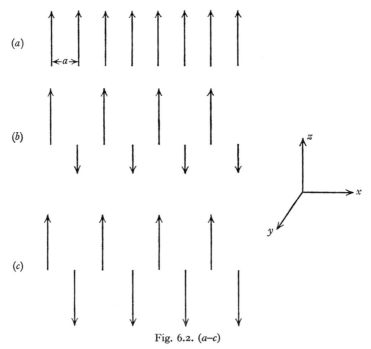

Fig. 6.2. (a–c)

Magnons

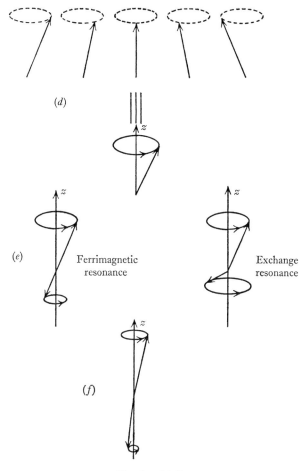

Fig. 6.2. (*d–f*)

(fig. 6.2(*b*)). In the antiferromagnetic arrangement the spin of the unpaired electron(s) of adjacent paramagnetic atoms are aligned antiparallel throughout the crystal so that there is a net zero magnetic moment for the whole crystal (fig. 6.2(*c*)).

In such ordered magnetic systems there is an exchange interaction between the spins on adjacent atoms which falls off exponentially with distance and introduces a coupling energy dependent upon the relative spin orientation on these atoms. As a result of the exponential fall-off of the coupling energy with distance only the coupling energy due to nearest neighbour coupling is usually considered. Figs. 6.2(*a*) to (*c*) represent

203

Other effects in vibrational energy region

the ground state of the system of coupled spins. Excited states where one spin in fig. 6.2(a) is reversed may be considered and such an excitation is represented by a situation where all spins share this reversal and all the spins precess in phase (fig. 6.2(d)), such an excitation wave being known as a spin wave. The quantum of energy in spin waves is known as the *magnon*. The elementary excitations of an ordered spin system are therefore known as magnons. The possible magnons in ferromagnetic (fig. 6.2(d)), ferrimagnetic (fig. 6.2(e)), and antiferromagnetic (fig. 6.2(f)) arrangements are illustrated in fig. 6.2.

The magnon energies for these various arrangements can be calculated, and Tinkham [747] has given a full description of such calculations. For the ferromagnetic case a dispersion relation analogous to the acoustic phonon results in the absence of an applied magnetic field, except that the energy at $K = 0$ is not zero because of the small anisotropy energy (which arises because anisotropic exchange favours the z direction), and the frequency of the $K = 0$ magnon can be written [747]

$$\nu = \frac{g\beta'}{h}(H_A + H_0), \qquad (6.5)$$

where β' is the Bohr magneton, H_A the anisotropy field and H_0 is the applied field in the z direction. When $H_0 = 0$ the $K = 0$ ferromagnetic magnon has an energy in the microwave and sometimes in the far infrared. For the ferrimagnetic case two types of magnon are possible. In one type, known as ferrimagnetic resonance, the spins of the two paramagnetic atoms are arranged antiparallel and collinear and precess uniformly as in the ferromagnetic case, so exhibiting a similar dispersion relation. In the other type, known as exchange resonance, the spins of the two paramagnetic atoms are arranged antiparallel but are not collinear and precess independently. This type of $K = 0$ magnon has an energy directly proportional to the exchange field [747]

$$\nu = \frac{g}{h}(\beta'_a M_a - \beta'_b M_b)(H_E + H_0), \qquad (6.6)$$

where H_E is the exchange field (it did not appear in (6.5) because in that case all the spins are collinear and thus precess uniformly), and M_a and M_b refer to the magnetic moments for the two types of atom a and b. When $H_0 = 0$ the $K = 0$ ferrimagnetic magnon has an energy in the infrared and a dispersion relation analogous to the optical phonon. For the antiferromagnetic case the spins are arranged antiparallel but not

Magnons

collinear and precess independently. The $K = 0$ magnon has an energy directly proportional to the geometric mean of the exchange and anisotropy fields [747]

$$\nu = \frac{g}{h}\beta'[H_0 \pm \sqrt{(2H_E H_A + H_A^2)}]. \tag{6.7}$$

When $H_0 = 0$ and $K = 0$ ferromagnetic magnon consists of two degenerate modes (the degeneracy is lifted when $H_0 \neq 0$) with an energy in the far infrared and a dispersion relation analogous to the acoustic phonon except that the energy at $K = 0$ is not zero.

6.4.1 One-magnon processes

The various magnon modes can be excited as fundamentals by the magnetic field of the electromagnetic radiation, conservation of wave vector leading to the excitation of $K = 0$ modes.

Magnons in ferromagnetic materials have been detected by vibrational spectroscopy. The rare earth metals Dy and Tb have a ferromagnetic phase at low temperature and give far infrared absorption due to $K = 0$ magnons (around $10\,\text{cm}^{-1}$) [748]. In Raman studies the small shifts (often of the order $1\,\text{cm}^{-1}$) make investigation difficult. Two different mechanisms for Raman scattering by spin waves have been proposed [5].

(i) Coupling of the radiation field directly with the magnons by magnetic–dipole interaction.

(ii) Coupling by spin-orbit interaction in a virtual intermediate state leading to Raman scattering.

Magnons in ferrimagnetic materials have also been detected by vibrational spectroscopy. Ytterbium iron garnet was the first ferrimagnet to have its far infrared absorption studied [749, 750]. Exchange resonance has been observed around $14\,\text{cm}^{-1}$, and ferrimagnetic resonance around $2\,\text{cm}^{-1}$, in such systems.

Magnons in antiferromagnetic materials have been more fully studied. The first far infrared absorption study [751] was made with FeF_2 (which becomes antiferromagnetic at low temperatures) which showed at $\approx 0\,°K$ an antiferromagnetic resonance mode at around $53\,\text{cm}^{-1}$. These modes are temperature dependent, the frequency falling with increasing temperature, since the anisotropy field is temperature dependent. Other studies have been compiled by Richards [752]. Solid polycrystalline oxygen has an antiferromagnetic α-phase between 0 and $23\,°K$ and this phase gives one-magnon absorption [753] at $27\,\text{cm}^{-1}$. Raman studies (thought to be by mechanism (ii) above) have shown one-magnon scat-

Other effects in vibrational energy region

tering in FeF_2 [754] at about 50 cm^{-1} in samples cooled below 50 °K, and in CoF_2 [755] at 37 cm^{-1} in samples cooled to 2 °K.

Examination of (6.5), (6.6) and (6.7) shows that one-magnon modes show a frequency dependence upon an applied magnetic field, which also lifts the degeneracy of the antiferromagnetic magnon modes.

6.4.2 Two-magnon processes

Two-magnon processes have been observed in vibrational spectroscopic studies of ferrimagnetic and antiferromagnetic materials with a frequency appreciably greater than twice the $K = 0$ magnon frequency. These two-magnon processes are excited by the electric field of the electromagnetic radiation, since their frequency is independent of applied magnetic fields in contrast to the one-magnon processes above.

Conservation of wave vector means that the two magnons have equal and opposite wave vectors (see (5.10) and (5.12)), but unlike the phonon density of states the magnon density of states does not have a maximum at $K = 0$, though it does have a maximum at the zone boundary. This means that the two-magnon density of states shows a maximum for two-magnon processes involving zone boundary magnons. Bearing in mind that the antiferromagnetic dispersion relation is similar to the acoustic phonon dispersion relation, the zone boundary magnons would be expected to have a much higher frequency than the $K = 0$ magnons, and therefore two-magnon processes have a frequency appreciably greater than twice the $K = 0$ magnon frequency. Loudon [756] has given a detailed discussion of selection rules for these two-magnon processes.

The two-magnon processes are excited by the electric field of the electromagnetic radiation because the mechanism of the process involves intermediate states where the virtual process is excited by the electric field. Three mechanisms have been proposed involving the following virtual processes.

(i) Electrical quadrupole–dipole interaction with the electric field of the radiation leads to virtual excitons which excite the two magnons via spin-orbit coupling [757, 758].

(ii) Non-diagonal exchange interactions between pairs of magnetic ions leading to exchange coupling with the electric field of the radiation [759].

(iii) An interaction scheme involving the excitation of virtual phonons by interaction with the electric field of the radiation, the virtual infrared active phonons leading to the two magnons [760].

Two-magnon processes have been observed in infrared absorption

Magnons

studies of FeF_2 [757] (154 cm^{-1}), and of MnF_2 [761] (106 cm^{-1}), which like FeF_2 becomes antiferromagnetic at low temperatures. Two-magnon bands have also been observed in NiF_2 [762] and CoF_2 [763].

Raman scattering by two-magnon processes has been observed for FeF_2 [754] (150 cm^{-1}), MnF_2 [764] (100 cm^{-1}), $RbMnF_3$ [756, 766] (132 cm^{-1}), $CsMnF_3$ [776] (93 cm^{-1}), $KNiF_3$ and K_2NiF_4 [777] (740 and 520 cm^{-1}), and NiO [778] (1560 cm^{-1}).

Raman scattering by localized two-magnon processes has also been observed. Thus the Raman scattering [767, 768] from MnF_2 doped with Ni^{2+} and doped with Fe^{2+} show the two magnons described above together with two-magnon bands due to localized antiferromagnetic magnons (a number of bands in the region 160–170 cm^{-1} (Ni^{2+} doping) and 140–170 cm^{-1} (Fe^{2+} doping)) together with a possible one-magnon localized antiferromagnet magnon at 26 cm^{-1} for Ni^{2+} doping. Localized two-magnon bands [769] have also been observed in Ni^{2+} doped $RbMnF_3$ (291 cm^{-1}) and $KMnF_3$ (313 cm^{-1}).

Raman scattering by two-magnon processes in ferrimagnetic substances has also been observed. $RbNiF_3$ becomes ferrimagnetic below 139 °K and shows two-magnon scattering at 500 cm^{-1} at 10 °K [770, 777].

6.4.3 *Magnon cluster excitations*

The magnons discussed above involved an excited state where one spin was reversed. Such an excited state obeys the selection rule $\Delta m = \pm 1$. Other excited states with more than one spin reversed correspond to n-fold multiple excitations and thus multiple-magnons would normally be disallowed by this selection rule. This selection rule can however be relaxed by exchange interaction and multiple-magnons may become allowed for certain substances. Magnon–magnon interactions in a multiple-magnon excitation can lead to a binding energy causing a multiple-magnon bound state [771] with an energy that is considerably less than n times the energy of the single-magnon process. Multiple-magnon bound states usually only occur near the Brillouin zone boundary and thus cannot be excited in a fundamental process. In the case of $CoCl_2 \cdot 2H_2O$ multiple-magnon bound states exist throughout the Brillouin zone including $K = 0$ and can thus be excited as fundamentals. In this case magnon–magnon interactions are so great (due to completely anisotropic exchange coupling [747]) that the energy of five-magnon bound states are approximately the same as the single-magnon state, though the bound states have five times the magnetic field dependence.

Other effects in vibrational energy region

Far infrared absorption studies [747, 772, 773] have identified fundamentals due to up to five-magnon bound states in the 30–60 cm^{-1} region. The effect of a magnetic field on these multiple-magnon bound states has been investigated; the field also causes changes in the nature of the crystal. Thus below 32 kOe the substance is antiferromagnetic, below 46 kOe it is ferrimagnetic and above 46 kOe it is ferromagnetic. There is also evidence for strong magnon–phonon interaction which leads to the excitation of the normally inactive phonon near 30 cm^{-1} (a phenomenon also observed in other systems [775]).

Appendix: character tables

A.1 THE C_s, C_i, AND C_n GROUPS

C_s	E	σ_h		
A'	1	1	(μ_x, μ_y)	$\alpha_{xx}\, \alpha_{yy}\, \alpha_{zz}\, \alpha_{xy}$
A''	1	-1	μ_z	$\alpha_{yz}\, \alpha_{zx}$

C_i	E	i		
A_g	1	1		α
A_u	1	-1	μ	

C_1	E		
A	1	μ	α

C_2	E	C_2		
A	1	1	μ_z	$\alpha_{xx}\, \alpha_{yy}\, \alpha_{zz}\, \alpha_{xy}$
B	1	-1	(μ_x, μ_y)	$\alpha_{yz}\, \alpha_{zx}$

C_3	E	C_3	C_3^2			$\epsilon = e^{\frac{2\pi i}{3}}$
A	1	1	1	μ_z	$\alpha_{xx}+\alpha_{yy}, \alpha_{zz}$	
E	$\begin{cases}1 \\ 1\end{cases}$	$\begin{matrix}\epsilon \\ \epsilon^*\end{matrix}$	$\begin{matrix}\epsilon^* \\ \epsilon\end{matrix}$	(μ_x, μ_y)	$(\alpha_{xx}-\alpha_{yy}, \alpha_{xy}), (\alpha_{yz}, \alpha_{zx})$	

C_4	E	C_4	C_2	C_4^3		
A	1	1	1	1	μ_z	$\alpha_{xx}+\alpha_{yy}, \alpha_{zz}$
B	1	-1	1	-1		$\alpha_{xx}-\alpha_{yy}, \alpha_{xy}$
E	$\begin{cases}1 \\ 1\end{cases}$	$\begin{matrix}i \\ -i\end{matrix}$	$\begin{matrix}-1 \\ -1\end{matrix}$	$\begin{matrix}-i \\ i\end{matrix}$	(μ_x, μ_y)	$(\alpha_{yz}, \alpha_{zx})$

C_5	E	C_5	C_5^2	C_5^3	C_5^4			$\epsilon = e^{\frac{2\pi i}{5}}$
A	1	1	1	1	1	μ_z	$\alpha_{xx}+\alpha_{yy}, \alpha_{zz}$	
E_1	$\begin{cases}1 \\ 1\end{cases}$	$\begin{matrix}\epsilon \\ \epsilon^*\end{matrix}$	$\begin{matrix}\epsilon^2 \\ \epsilon^{2*}\end{matrix}$	$\begin{matrix}\epsilon^{2*} \\ \epsilon^2\end{matrix}$	$\begin{matrix}\epsilon^* \\ \epsilon\end{matrix}$	(μ_x, μ_y)	$(\alpha_{yz}, \alpha_{zx})$	
E_2	$\begin{cases}1 \\ 1\end{cases}$	$\begin{matrix}\epsilon^2 \\ \epsilon^{2*}\end{matrix}$	$\begin{matrix}\epsilon^* \\ \epsilon\end{matrix}$	$\begin{matrix}\epsilon \\ \epsilon^*\end{matrix}$	$\begin{matrix}\epsilon^{2*} \\ \epsilon^2\end{matrix}$		$(\alpha_{xx}-\alpha_{yy}, \alpha_{xy})$	

C_6	E	C_6	C_3	C_2	C_3^2	C_6^5			$\epsilon = e^{\frac{2\pi i}{6}}$
A	1	1	1	1	1	1	μ_z	$\alpha_{xx}+\alpha_{yy}, \alpha_{zz}$	
B	1	-1	1	-1	1	-1			
E_1	$\begin{cases}1 \\ 1\end{cases}$	$\begin{matrix}\epsilon \\ \epsilon^*\end{matrix}$	$\begin{matrix}-\epsilon^* \\ -\epsilon\end{matrix}$	$\begin{matrix}-1 \\ -1\end{matrix}$	$\begin{matrix}-\epsilon \\ -\epsilon^*\end{matrix}$	$\begin{matrix}\epsilon^* \\ \epsilon\end{matrix}$	(μ_x, μ_y)	$(\alpha_{yz}, \alpha_{zx})$	
E_2	$\begin{cases}1 \\ 1\end{cases}$	$\begin{matrix}-\epsilon^* \\ -\epsilon\end{matrix}$	$\begin{matrix}-\epsilon \\ -\epsilon^*\end{matrix}$	$\begin{matrix}1 \\ 1\end{matrix}$	$\begin{matrix}-\epsilon^* \\ -\epsilon\end{matrix}$	$\begin{matrix}-\epsilon \\ -\epsilon^*\end{matrix}$		$(\alpha_{xx}-\alpha_{yy}, \alpha_{xy})$	

Appendix

A.2 THE D_n GROUPS

$D_2 = V$	E	$C_2(z)$	$C_2(y)$	$C_2(x)$		
A	1	1	1	1		$\alpha_{xx}, \alpha_{yy}, \alpha_{zz}$
B_1	1	1	-1	-1	μ_z	α_{xy}
B_2	1	-1	1	-1	μ_y	α_{zx}
B_3	1	-1	-1	1	μ_x	α_{yz}

D_3	E	$2C_3$	$3C_2$		
A_1	1	1	1		$\alpha_{xx}+\alpha_{yy}, \alpha_{zz}$
A_2	1	1	-1	μ_z	
E	2	-1	0	(μ_x, μ_y)	$(\alpha_{xx}-\alpha_{yy}, \alpha_{xy}); (\alpha_{yz}, \alpha_{zx})$

D_4	E	$2C_4$	$C_4^2=C_2$	$2C_2'$	$2C_2''$		
A_1	1	1	1	1	1		$\alpha_{xx}+\alpha_{yy}, \alpha_{zz}$
A_2	1	1	1	-1	-1	μ_z	
B_1	1	-1	1	1	-1		$\alpha_{xx}-\alpha_{yy}$
B_2	1	-1	1	-1	1		α_{xy}
E	2	0	-2	0	0	(μ_x, μ_y)	$(\alpha_{yz}, \alpha_{zx})$

D_5	E	$2C_5$	$2C_5^2$	$5C_2$		
A_1	1	1	1	1		$\alpha_{xx}+\alpha_{yy}, \alpha_{zz}$
A_2	1	1	1	-1	μ_z	
E_1	2	$2\cos 72°$	$2\cos 144°$	0	(μ_x, μ_y)	$(\alpha_{yz}, \alpha_{zx})$
E_2	2	$2\cos 144°$	$2\cos 72°$	0		$(\alpha_{xx}-\alpha_{yy}, \alpha_{xy})$

D_6	E	$2C_6$	$2C_3$	C_2	$3C_2'$	$3C_2''$		
A_1	1	1	1	1	1	1		$\alpha_{xx}+\alpha_{yy}, \alpha_{zz}$
A_2	1	1	1	1	-1	-1	μ_z	
B_1	1	-1	1	-1	1	-1		
B_2	1	-1	1	-1	-1	1		
E_1	2	1	-1	-2	0	0	(μ_x, μ_y)	$(\alpha_{yz}, \alpha_{zx})$
E_2	2	-1	-1	2	0	0		$(\alpha_{xx}-\alpha_{yy}, \alpha_{xy})$

A.3 THE C_{nv} GROUPS

C_{2v}	E	C_2	$\sigma_v(zx)$	$\sigma_v(yz)$		
A_1	1	1	1	1	μ_z	$\alpha_{xx}, \alpha_{yy}, \alpha_{zz}$
A_2	1	1	-1	-1		α_{xy}
B_1	1	-1	1	-1	μ_x	α_{zx}
B_2	1	-1	-1	1	μ_y	α_{yz}

C_{3v}	E	$2C_3$	$3\sigma_v$		
A_1	1	1	1	μ_z	$\alpha_{xx}+\alpha_{yy}, \alpha_{zz}$
A_2	1	1	-1		
E	2	-1	0	(μ_x, μ_y)	$(\alpha_{xx}-\alpha_{yy}, \alpha_{xy}); (\alpha_{yz}, \alpha_{zx})$

Character tables

C_{4v}	E	$2C_4$	C_2	$2\sigma_v$	$2\sigma_d$		
A_1	1	1	1	1	1	μ_z	$\alpha_{xx}+\alpha_{yy}, \alpha_{zz}$
A_2	1	1	1	-1	-1		
B_1	1	-1	1	1	-1		$\alpha_{xx}-\alpha_{yy}$
B_2	1	-1	1	-1	1		α_{xy}
E	2	0	-2	0	0	(μ_x, μ_y)	$(\alpha_{yz}, \alpha_{zx})$

C_{5v}	E	$2C_5$	$2C_5^2$	$5\sigma_v$		
A_1	1	1	1	1	μ_z	$\alpha_{xx}+\alpha_{yy}, \alpha_{zz}$
A_2	1	1	1	-1		
E_1	2	$2\cos 72°$	$2\cos 144°$	0	(μ_x, μ_z)	$(\alpha_{yz}, \alpha_{zx})$
E_2	2	$2\cos 144°$	$2\cos 72°$	0		$(\alpha_{xx}-\alpha_{yy}, \alpha_{xy})$

C_{6v}	E	$2C_6$	$2C_3$	C_2	$3\sigma_v$	$3\sigma_d$		
A_1	1	1	1	1	1	1	μ_z	$\alpha_{xx}+\alpha_{yy}, \alpha_{zz}$
A_2	1	1	1	1	-1	-1		
B_1	1	-1	1	-1	1	-1		
B_2	1	-1	1	-1	-1	1		
E_1	2	1	-1	-2	0	0	(μ_x, μ_y)	$(\alpha_{yz}, \alpha_{zx})$
E_2	2	-1	-1	2	0	0		$(\alpha_{xx}-\alpha_{yy}, \alpha_{xy})$

A.4 THE C_{nh} GROUPS

C_{2h}	E	C_2	i	σ_h		
A_g	1	1	1	1		$\alpha_{xx}, \alpha_{yy}, \alpha_{zz}, \alpha_{xy}$
B_g	1	-1	1	-1		α_{yz}, α_{zx}
A_u	1	1	-1	-1	μ_z	
B_u	1	-1	-1	1	(μ_x, μ_y)	

C_{3h}	E	C_3	C_3^2	σ_h	S_3	S_3^5		$\epsilon = e^{\frac{2\pi i}{3}}$
A'	1	1	1	1	1	1		$\alpha_{xx}+\alpha_{yy}, \alpha_{zz}$
E'	$\begin{cases}1\\1\end{cases}$	$\begin{matrix}\epsilon\\ \epsilon^*\end{matrix}$	$\begin{matrix}\epsilon^*\\ \epsilon\end{matrix}$	$\begin{matrix}1\\1\end{matrix}$	$\begin{matrix}\epsilon\\ \epsilon^*\end{matrix}$	$\begin{matrix}\epsilon^*\\ \epsilon\end{matrix}$	(μ_x, μ_y)	$(\alpha_{xx}-\alpha_{yy}, \alpha_{xy})$
A''	1	1	1	-1	-1	-1	μ_z	
E''	$\begin{cases}1\\1\end{cases}$	$\begin{matrix}\epsilon\\ \epsilon^*\end{matrix}$	$\begin{matrix}\epsilon^*\\ \epsilon\end{matrix}$	$\begin{matrix}-1\\-1\end{matrix}$	$\begin{matrix}-\epsilon\\ -\epsilon^*\end{matrix}$	$\begin{matrix}-\epsilon^*\\ -\epsilon\end{matrix}$		$(\alpha_{yz}, \alpha_{zx})$

C_{4h}	E	C_4	C_2	C_4^3	i	S_4^3	σ_h	S_4		
A_g	1	1	1	1	1	1	1	1		$\alpha_{xx}+\alpha_{yy}, \alpha_{zz}$
B_g	1	-1	1	-1	1	-1	1	-1		$\alpha_{xx}-\alpha_{yy}, \alpha_{xy}$
E_g	$\begin{cases}1\\1\end{cases}$	$\begin{matrix}i\\-i\end{matrix}$	$\begin{matrix}-1\\-1\end{matrix}$	$\begin{matrix}-i\\i\end{matrix}$	$\begin{matrix}1\\1\end{matrix}$	$\begin{matrix}i\\-i\end{matrix}$	$\begin{matrix}-1\\-1\end{matrix}$	$\begin{matrix}-i\\i\end{matrix}$		$(\alpha_{yz}, \alpha_{zx})$
A_u	1	1	1	1	-1	-1	-1	-1	μ_z	
B_u	1	-1	1	-1	-1	1	-1	1		
E_u	$\begin{cases}1\\1\end{cases}$	$\begin{matrix}i\\-i\end{matrix}$	$\begin{matrix}-1\\-1\end{matrix}$	$\begin{matrix}-i\\i\end{matrix}$	$\begin{matrix}-1\\-1\end{matrix}$	$\begin{matrix}-i\\i\end{matrix}$	$\begin{matrix}1\\1\end{matrix}$	$\begin{matrix}i\\-i\end{matrix}$	(μ_x, μ_y)	

Appendix

C_{5h}	E	C_5	C_5^2	C_5^3	C_5^4	σ_h	S_5	S_5^7	S_5^3	S_5^9		$\epsilon = e^{\frac{2\pi i}{5}}$
A'	1	1	1	1	1	1	1	1	1	1		$\alpha_{xx}+\alpha_{yy}, \alpha_{zz}$
E_1'	1	ϵ	ϵ^2	ϵ^{2*}	ϵ^*	1	ϵ	ϵ^2	ϵ^{2*}	ϵ^*	(μ_x, μ_y)	
	1	ϵ^*	ϵ^{2*}	ϵ^2	ϵ	1	ϵ^*	ϵ^{2*}	ϵ^2	ϵ		
E_2'	1	ϵ^2	ϵ^*	ϵ	ϵ^{2*}	1	ϵ^2	ϵ^*	ϵ	ϵ^{2*}		$(\alpha_{xx}-\alpha_{yy}, \alpha_{xy})$
	1	ϵ^{2*}	ϵ	ϵ^*	ϵ^2	1	ϵ^{2*}	ϵ	ϵ^*	ϵ^2		
A''	1	1	1	1	1	-1	-1	-1	-1	-1	μ_z	
E_1''	1	ϵ	ϵ^2	ϵ^{2*}	ϵ^*	-1	$-\epsilon$	$-\epsilon^2$	$-\epsilon^{2*}$	$-\epsilon^*$		α_{yz}, α_{zx}
	1	ϵ^*	ϵ^{2*}	ϵ^2	ϵ	-1	$-\epsilon^*$	$-\epsilon^{2*}$	$-\epsilon^2$	$-\epsilon$		
E_2''	1	ϵ^2	ϵ^*	ϵ	ϵ^{2*}	-1	$-\epsilon^2$	$-\epsilon^*$	$-\epsilon$	$-\epsilon^{2*}$		
	1	ϵ^{2*}	ϵ	ϵ^*	ϵ^2	-1	$-\epsilon^{2*}$	$-\epsilon$	$-\epsilon^*$	$-\epsilon^2$		

C_{6h}	E	C_6	C_3	C_2	C_3^2	C_6^5	i	S_3^5	S_6^5	σ_h	S_6	S_3		$\epsilon = e^{\frac{2\pi i}{6}}$
A_g	1	1	1	1	1	1	1	1	1	1	1	1		$\alpha_{xx}+\alpha_{yy}, \alpha_{zz}$
B_g	1	-1	1	-1	1	-1	1	-1	1	-1	1	-1		
E_{1g}	1	ϵ	$-\epsilon^*$	-1	$-\epsilon$	ϵ^*	1	ϵ	$-\epsilon^*$	-1	$-\epsilon$	ϵ^*		$(\alpha_{yz}, \alpha_{zx})$
	1	ϵ^*	$-\epsilon$	-1	$-\epsilon^*$	ϵ	1	ϵ^*	$-\epsilon$	-1	$-\epsilon^*$	ϵ		
E_{2g}	1	$-\epsilon^*$	$-\epsilon$	1	$-\epsilon^*$	$-\epsilon$	1	$-\epsilon^*$	$-\epsilon$	1	$-\epsilon^*$	$-\epsilon$		$(\alpha_{xx}-\alpha_{yy}, \alpha_{xy})$
	1	$-\epsilon$	$-\epsilon^*$	1	$-\epsilon$	$-\epsilon^*$	1	$-\epsilon$	$-\epsilon^*$	1	$-\epsilon$	$-\epsilon^*$		
A_u	1	1	1	1	1	1	-1	-1	-1	-1	-1	-1	μ_z	
B_u	1	-1	1	-1	1	-1	-1	1	-1	1	-1	1		
E_{1u}	1	ϵ	$-\epsilon^*$	-1	$-\epsilon$	ϵ^*	-1	$-\epsilon$	ϵ^*	1	ϵ	$-\epsilon^*$	(μ_x, μ_y)	
	1	ϵ^*	$-\epsilon$	-1	$-\epsilon^*$	ϵ	-1	$-\epsilon^*$	ϵ	1	ϵ^*	$-\epsilon$		
E_{2u}	1	$-\epsilon^*$	$-\epsilon$	1	$-\epsilon^*$	$-\epsilon$	-1	ϵ^*	ϵ	-1	ϵ^*	ϵ		
	1	$-\epsilon$	$-\epsilon^*$	1	$-\epsilon$	$-\epsilon^*$	-1	ϵ	ϵ^*	-1	ϵ	ϵ^*		

A. 5 THE D_{nh} GROUPS

$D_{2h} = V_h$	E	$C_2(z)$	$C_2(y)$	$C_2(x)$	i	$\sigma(xy)$	$\sigma(zx)$	$\sigma(yz)$		
A_g	1	1	1	1	1	1	1	1		$\alpha_{xx}, \alpha_{yy}, \alpha_{zz}$
B_{1g}	1	1	-1	-1	1	1	-1	-1		α_{xy}
B_{2g}	1	-1	1	-1	1	-1	1	-1		α_{zx}
B_{3g}	1	-1	-1	1	1	-1	-1	1		α_{yz}
A_u	1	1	1	1	-1	-1	-1	-1		
B_{1u}	1	1	-1	-1	-1	-1	1	1	μ_z	
B_{2u}	1	-1	1	-1	-1	1	-1	1	μ_y	
B_{3u}	1	-1	-1	1	-1	1	1	-1	μ_x	

D_{3h}	E	$2C_3$	$3C_2$	σ_h	$2S_3$	$3\sigma_v$		
A_1'	1	1	1	1	1	1		$\alpha_{xx}+\alpha_{yy}, \alpha_{zz}$
A_2'	1	1	-1	1	1	-1		
E'	2	-1	0	2	-1	0	(μ_x, μ_y)	$(\alpha_{xx}-\alpha_{yy}, \alpha_{xy})$
A_1''	1	1	1	-1	-1	-1		
A_2''	1	1	-1	-1	-1	1	μ_z	
E''	2	-1	0	-2	1	0		$(\alpha_{yz}, \alpha_{zx})$

Character tables

D_{4h}	E	$2C_4$	C_2	$2C_2'$	$2C_2''$	i	$2S_4$	σ_h	$2\sigma_v$	$2\sigma_h$	
A_{1g}	1	1	1	1	1	1	1	1	1	1	$\alpha_{xx}+\alpha_{yy}, \alpha_{zz}$
A_{2g}	1	1	1	−1	−1	1	1	1	−1	−1	
B_{1g}	1	−1	1	1	−1	1	−1	1	1	−1	$\alpha_{xx}-\alpha_{yy}$
B_{2g}	1	−1	1	−1	1	1	−1	1	−1	1	α_{xy}
E_g	2	0	−2	0	0	2	0	−2	0	0	$(\alpha_{yz}, \alpha_{zx})$
A_{1u}	1	1	1	1	1	−1	−1	−1	−1	−1	
A_{2u}	1	1	1	−1	−1	−1	−1	−1	1	1	μ_z
B_{1u}	1	−1	1	1	−1	−1	1	−1	−1	1	
B_{2u}	1	−1	1	−1	1	−1	1	−1	1	−1	
E_u	2	0	−2	0	0	−2	0	2	0	0	(μ_x, μ_y)

D_{5h}	E	$2C_5$	$2C_5^2$	$5C_2$	σ_h	$2S_5$	$2S_5^3$	$5\sigma_v$	
A_1'	1	1	1	1	1	1	1	1	$\alpha_{xx}+\alpha_{yy}, \alpha_{zz}$
A_2'	1	1	1	−1	1	1	1	−1	
E_1'	2	2 cos 72°	2 cos 144°	0	2	2 cos 72°	2 cos 144°	0	(μ_x, μ_y)
E_2'	2	2 cos 144°	2 cos 72°	0	2	2 cos 144°	2 cos 72°	0	$(\alpha_{xx}-\alpha_{yy}, \alpha_{xy})$
A_1''	1	1	1	1	−1	−1	−1	−1	
A_2''	1	1	1	−1	−1	−1	−1	1	μ_z
E_1''	2	2 cos 72°	2 cos 144°	0	−2	−2 cos 72°	−2 cos 144°	0	$(\alpha_{yz}, \alpha_{zx})$
E_2''	2	2 cos 144°	2 cos 72°	0	−2	−2 cos 144°	−2 cos 72°	0	

D_{6h}	E	$2C_6$	$2C_3$	C_2	$3C_2'$	$3C_2''$	i	$2S_3$	$2S_6$	σ_h	$3\sigma_d$	$3\sigma_v$	
A_{1g}	1	1	1	1	1	1	1	1	1	1	1	1	$\alpha_{xx}+\alpha_{yy}, \alpha_{zz}$
A_{2g}	1	1	1	1	−1	−1	1	1	1	1	−1	−1	
B_{1g}	1	−1	1	−1	1	−1	1	−1	1	−1	1	−1	
B_{2g}	1	−1	1	−1	−1	1	1	−1	1	−1	−1	1	
E_{1g}	2	1	−1	−2	0	0	2	1	−1	−2	0	0	$(\alpha_{yz}, \alpha_{zx})$
E_{2g}	2	−1	−1	2	0	0	2	−1	−1	2	0	0	$(\alpha_{xx}-\alpha_{yy}, \alpha_{xy})$
A_{1u}	1	1	1	1	1	1	−1	−1	−1	−1	−1	−1	
A_{2u}	1	1	1	1	−1	−1	−1	−1	−1	−1	1	1	μ_z
B_{1u}	1	−1	1	−1	1	−1	−1	1	−1	1	−1	1	
B_{2u}	1	−1	1	−1	−1	1	−1	1	−1	1	1	−1	
E_{1u}	2	1	−1	−2	0	0	−2	−1	1	2	0	0	(μ_x, μ_y)
E_{2u}	2	−1	−1	2	0	0	−2	1	1	−2	0	0	

A.6 THE D_{nd} GROUPS

$D_{2d} = V_d$	E	$2S_4$	C_2	$2C_2'$	$2\sigma_d$		
A_1	1	1	1	1	1		$\alpha_{xx}+\alpha_{yy}, \alpha_{zz}$
A_2	1	1	1	−1	−1		
B_1	1	−1	1	1	−1		$\alpha_{xx}-\alpha_{yy}$
B_2	1	−1	1	−1	1	μ_z	α_{xy}
E	2	0	−2	0	0	(μ_x, μ_y)	$(\alpha_{yz}, \alpha_{zx})$

213

Appendix

D_{3d}	E	$2C_3$	$3C_2$	i	$2S_6$	$3\sigma_d$		
A_{1g}	1	1	1	1	1	1		$\alpha_{xx}+\alpha_{yy}, \alpha_{zz}$
A_{2g}	1	1	−1	1	1	−1		
E_g	2	−1	0	2	−1	0		$(\alpha_{xx}-\alpha_{yy}, \alpha_{xy}), (\alpha_{yz}, \alpha_{zx})$
A_{1u}	1	1	1	−1	−1	−1		
A_{2u}	1	1	−1	−1	−1	1	μ_z	
E_u	2	−1	0	−2	1	0	(μ_x, μ_y)	

D_{4d}	E	$2S_8$	$2C_4$	$2S_8^3$	C_2	$4C_2'$	$4\sigma_d$		
A_1	1	1	1	1	1	1	1		$\alpha_{xx}+\alpha_{yy}, \alpha_{zz}$
A_2	1	1	1	1	1	−1	−1		
B_1	1	−1	1	−1	1	1	−1		
B_2	1	−1	1	−1	1	−1	1	μ_z	
E_1	2	$\sqrt{2}$	0	$-\sqrt{2}$	−2	0	0	(μ_x, μ_y)	
E_2	2	0	−2	0	2	0	0		$(\alpha_{xx}-\alpha_{yy}, \alpha_{xy})$
E_3	2	$-\sqrt{2}$	0	$\sqrt{2}$	−2	0	0		$(\alpha_{yz}, \alpha_{zx})$

D_{5d}	E	$2C_5$	$2C_5^2$	$5C_2$	i	$2S_{10}^3$	$2S_{10}$	$5\sigma_d$		
A_{1g}	1	1	1	1	1	1	1	1		$\alpha_{xx}+\alpha_{yy}, \alpha_{zz}$
A_{2g}	1	1	1	−1	1	1	1	−1		
E_{1g}	2	$2\cos 72°$	$2\cos 144°$	0	2	$2\cos 72°$	$2\cos 144°$	0		$(\alpha_{yz}, \alpha_{zx})$
E_{2g}	2	$2\cos 144°$	$2\cos 72°$	0	2	$2\cos 144°$	$2\cos 72°$	0		$(\alpha_{xx}-\alpha_{yy}, \alpha_{xy})$
A_{1u}	1	1	1	1	−1	−1	−1	−1		
A_{2u}	1	1	1	−1	−1	−1	−1	1	μ_z	
E_{1u}	2	$2\cos 72°$	$2\cos 144°$	0	−2	$-2\cos 72°$	$-2\cos 144°$	0	(μ_x, μ_y)	
E_{2u}	2	$2\cos 144°$	$2\cos 72°$	0	−2	$-2\cos 144°$	$-2\cos 72°$	0		

D_{6d}	E	$2S_{12}$	$2C_6$	$2S_4$	$2C_3$	$2S_{12}^5$	C_2	$6C_2'$	$6\sigma_d$		
A_1	1	1	1	1	1	1	1	1	1		$\alpha_{xx}+\alpha_{yy}, \alpha_{zz}$
A_2	1	1	1	1	1	1	1	−1	−1		
B_1	1	−1	1	−1	1	−1	1	1	−1		
B_2	1	−1	1	−1	1	−1	1	−1	1	μ_z	
E_1	2	$\sqrt{3}$	1	0	−1	$-\sqrt{3}$	−2	0	0	(μ_x, μ_y)	
E_2	2	1	−1	−2	−1	1	2	0	0		$(\alpha_{xx}-\alpha_{yy}, \alpha_{xy})$
E_3	2	0	−2	0	2	0	−2	0	0		
E_4	2	−1	−1	2	−1	−1	2	0	0		
E_5	2	$-\sqrt{3}$	1	0	−1	$\sqrt{3}$	−2	0	0		$(\alpha_{yz}, \alpha_{zx})$

A.7 THE S_n GROUPS

S_4	E	S_4	C_2	S_4^3			
A	1	1	1	1		$\alpha_{xx}+\alpha_{yy}, \alpha_{zz}$	
B	1	−1	1	−1	μ_z	$\alpha_{xx}-\alpha_{yy}, \alpha_{xy}$	
E	$\begin{Bmatrix} 1 \\ 1 \end{Bmatrix}$	$\begin{Bmatrix} i \\ -i \end{Bmatrix}$	$\begin{Bmatrix} -1 \\ -1 \end{Bmatrix}$	$\begin{Bmatrix} -i \\ i \end{Bmatrix}$	(μ_x, μ_y)	$(\alpha_{yz}, \alpha_{zx})$	

Character tables

S_6	E	C_3	C_3^2	i	S_6^5	S_6		$S_6 = C_3 \times i$ $\quad \epsilon = e^{\frac{2\pi i}{3}}$
A_g	1	1	1	1	1	1		$\alpha_{xx}+\alpha_{yy}, \alpha_{zz}$
E_g	$\begin{cases}1\\1\end{cases}$	ϵ ϵ^*	ϵ^* ϵ	1 1	ϵ ϵ^*	ϵ^* ϵ $\}$		$(\alpha_{xx}-\alpha_{yy}, \alpha_{xy}); (\alpha_{yz}, \alpha_{zx})$
A_u	1	1	1	-1	-1	-1	μ_z	
E_u	$\begin{cases}1\\1\end{cases}$	ϵ ϵ^*	ϵ^* ϵ	-1 -1	$-\epsilon$ $-\epsilon^*$	$-\epsilon^*$ $-\epsilon$ $\}$	(μ_x, μ_y)	

S_8	E	S_8	C_4	S_8^3	C_2	S_8^5	C_4^3	S_8^7		$\epsilon = e^{\frac{2\pi i}{8}}$
A	1	1	1	1	1	1	1	1		$(\alpha_{xx}+\alpha_{yy}, \alpha_{zz}$
B	1	-1	1	-1	1	-1	1	-1	μ_z	
E_1	$\begin{cases}1\\1\end{cases}$	ϵ ϵ^*	i $-i$	$-\epsilon^*$ $-\epsilon$	-1 -1	$-\epsilon$ $-\epsilon^*$	$-i$ i	ϵ^* ϵ $\}$	(μ_x, μ_y)	
E_2	$\begin{cases}1\\1\end{cases}$	i $-i$	-1 -1	$-i$ i	1 1	i $-i$	-1 -1	$-i$ i $\}$		$(\alpha_{xx}-\alpha_{yy}, \alpha_{xy})$
E_3	$\begin{cases}1\\1\end{cases}$	$-\epsilon^*$ $-\epsilon$	$-i$ i	ϵ ϵ^*	-1 -1	ϵ^* ϵ	i $-i$	$-\epsilon$ $-\epsilon^*$ $\}$		$(\alpha_{yz}, \alpha_{zx})$

A.8 THE CUBIC GROUPS

T	E	$4C_3$	$4C_3^2$	$3C_2$		$\epsilon = e^{\frac{2\pi i}{3}}$
A	1	1	1	1		$\alpha_{xx}+\alpha_{yy}+\alpha_{zz}$
E	$\begin{cases}1\\1\end{cases}$	ϵ ϵ^*	ϵ^* ϵ	1 1 $\}$		$(\alpha_{xx}+\alpha_{yy}-2\alpha_{zz}, \alpha_{xx}-\alpha_{yy})$
F	3	0	0	-1	μ	$(\alpha_{xy}, \alpha_{yz}, \alpha_{zx})$

T_h	E	$4C_3$	$4C_3^2$	$3C_2$	i	$4S_6^5$	$4S_6$	$3\sigma_h$		$\epsilon = e^{\frac{2\pi i}{3}}$
A_g	1	1	1	1	1	1	1	1		$\alpha_{xx}+\alpha_{yy}+\alpha_{zz}$
A_u	1	1	1	1	-1	-1	-1	-1		
E_g	$\begin{cases}1\\1\end{cases}$	ϵ ϵ^*	ϵ^* ϵ	1 1	1 1	ϵ ϵ^*	ϵ^* ϵ	1 1		$(\alpha_{xx}+\alpha_{yy}-2\alpha_{zz}, \alpha_{xx}-\alpha_{yy})$
E_u	$\begin{cases}1\\1\end{cases}$	ϵ ϵ^*	ϵ^* ϵ	1 1	-1 -1	$-\epsilon$ $-\epsilon^*$	$-\epsilon^*$ $-\epsilon$	-1 -1		
F_g	3	0	0	-1	3	0	0	-1		$(\alpha_{xy}, \alpha_{yz}, \alpha_{zx})$
F_u	3	0	0	-1	-3	0	0	1	μ	

T_d	E	$8C_3$	$3C_2$	$6S_4$	$6\sigma_d$		
A_1	1	1	1	1	1		$\alpha_{xx}+\alpha_{yy}+\alpha_{zz}$
A_2	1	1	1	-1	-1		
E	2	-1	2	0	0		$(\alpha_{xx}+\alpha_{yy}-2\alpha_{zz}, \alpha_{xx}-\alpha_{yy})$
F_1	3	0	-1	1	-1		
F_2	3	0	-1	-1	1	μ	$(\alpha_{xy}, \alpha_{yz}, \alpha_{zx})$

Appendix

O	E	$8C_3$	$3C_2$	$6C_4$	$6C_2'$		
A_1	1	1	1	1	1		$\alpha_{xx}+\alpha_{yy}+\alpha_{zz}$
A_2	1	1	1	−1	−1		
E	2	−1	2	0	0		$(\alpha_{xx}+\alpha_{yy}-2\alpha_{zz}, \alpha_{xx}-\alpha_{yy})$
F_1	3	0	−1	1	−1	μ	
F_2	3	0	−1	−1	1		$(\alpha_{xy}, \alpha_{yz}, \alpha_{zx})$

O_h	E	$8C_3$	$6C_2$	$6C_4$	$3C_2\ (=C_4^2)$	i	$6S_4$	$8S_6$	$3\sigma_h$	$6\sigma_d$		
A_{1g}	1	1	1	1	1	1	1	1	1	1		$\alpha_{xx}+\alpha_{yy}+\alpha_{zz}$
A_{2g}	1	1	−1	−1	1	1	−1	1	1	−1		
E_g	2	−1	0	0	2	2	0	−1	2	0		$(\alpha_{xx}+\alpha_{yy}-2\alpha_{zz},$ $\alpha_{xx}-\alpha_{yy})$
T_{1g}	3	0	−1	1	−1	3	1	0	−1	−1		
T_{2g}	3	0	1	−1	−1	3	−1	0	−1	1		$(\alpha_{xy}, \alpha_{yz}, \alpha_{zx})$
A_{1u}	1	1	1	1	1	−1	−1	−1	−1	−1		
A_{2u}	1	1	−1	−1	1	−1	1	−1	−1	1		
E_u	2	−1	0	0	2	−2	0	1	−2	0		
T_{1u}	3	0	−1	1	−1	−3	−1	0	1	1	μ	
T_{2u}	3	0	1	−1	−1	−3	1	0	1	−1		

A. 9 THE GROUPS $C_{\infty v}$ AND $D_{\infty h}$

$C_{\infty v}$	E	$2C_\infty^\phi$	...	$\infty\sigma_d$		
$A_1 \equiv \Sigma^+$	1	1	...	1	μ_z	$\alpha_{xx}+\alpha_{yy}, \alpha_{zz}$
$A_2 \equiv \Sigma^-$	1	1	...	−1		
$E_1 \equiv \Pi$	2	$2\cos\phi$	...	0	(μ_x, μ_y)	$(\alpha_{yz}, \alpha_{zx})$
$E_2 \equiv \Delta$	2	$2\cos 2\phi$	...	0		$(\alpha_{xx}-\alpha_{yy}, \alpha_{xy})$
$E_3 \equiv \Phi$	2	$2\cos 3\phi$	...	0		
...	...	...	...	...		

$D_{\infty h}$	E	$2C_\infty^\phi$	...	$\infty\sigma_v$	i	$2S_\infty^\phi$	...	∞C_2		
Σ_g^+	1	1	...	1	1	1	...	1		$\alpha_{xx}+\alpha_{yy}, \alpha_{zz}$
Σ_g^-	1	1	...	−1	1	1	...	−1		
Π_g	2	$2\cos\phi$	...	0	2	$-2\cos\phi$	...	0		$(\alpha_{yz}, \alpha_{zx})$
Δ_g	2	$2\cos 2\phi$	...	0	2	$2\cos 2\phi$	...	0		$(\alpha_{xx}-\alpha_{yy}, \alpha_{yx})$
...	...	...	...	...	...	...	...	...		
Σ_u^+	1	1	...	1	−1	−1	...	−1	μ_z	
Σ_u^-	1	1	...	−1	−1	−1	...	1		
Π_u	2	$2\cos\phi$	...	0	−2	$2\cos\phi$	...	0	(μ_x, μ_y)	
Δ_u	2	$2\cos 2\phi$	...	0	−2	$-2\cos 2\phi$	...	0		
...	...	...	...	...	...	...				

References

1 D. F. Hornig, *J. chem. phys.* **16**, 1063 (1948).
2 *Phonons in perfect lattices and in lattices with point imperfections*, edited by R. W. H. Stevenson, Oliver and Boyd, Edinburgh and London (1966).
3 D. H. Martin, *Adv. phys.* **14**, 39 (1965).
4 W. Vedder and D. F. Hornig, in *Advances in Spectroscopy* vol. 2, p. 189, edited by H. W. Thompson, Wiley–Interscience, New York (1961).
5 R. Loudon, *Adv. phys.* **13**, 423 (1964).
6 S. S. Mitra, *Solid state phys.* **13**, 66 (1962).
7 J. P. Mathieu, *Spectres de Vibration et Symétrie des Molécules et des Cristaux*, Herman et Cie, Paris (1945).
8 J. P. Mathieu, *J. phys. radium* **16**, 219 (1955).
9 J. P. Mathieu, *Memoires de la Societe Royale des sciences de Liege* **20**, 103 (1970).
10 A. Hadni, *Memoires de la Societe Royale des sciences de Liege* **20**, 33 (1970).
11 S. S. Mitra and P. J. Gielisse, in *Progress in infra-red spectroscopy* vol. 2, p. 47, edited by H. A. Szymanski, Plenum Press, New York (1964).
12 S. S. Mitra, in *Optical Properties of Solids* chapter 14, edited by S. Nudelman and S. S. Mitra, Plenum Press, New York (1969).
13 A. Finch, P. N. Gates, K. Radcliffe, F. N. Dickson and F. F. Bentley, *Chemical applications of far infra-red spectroscopy* chapter 8, Academic Press, London (1970).
14 A. E. Hughes, *Contemporary phys.* **12**, 231 (1971).
15 T. R. Gilson and P. J. Hendra, *Laser Raman Spectroscopy*, Wiley–Interscience, London (1970).
16 G. R. Wilkinson, in *Molecular Dynamics and Structure of Solids* p. 77. N.B.S. special publication 301 (1969).
17 E. Gross and M. Vuks, *Nature* **135**, 100 (1935).
18 J. P. Devlin, P. C. Li and G. Pollard, *J. chem. phys.* **52**, 2267 (1970).
19 V. Wagner, *Z. Physik* **224**, 353 (1969).
20 C. A. Angell, J. Wong and W. F. Edgell, *J. chem. phys.* **51**, 4519 (1969).
21 A. D. B. Woods and R. A. Cowley, *Phys. rev. lett.* **24**, 646 (1970).
22 A. V. R. Warrier and S. Khim, *J. chem. phys.* **52**, 4316 (1970).
23 G. H. Wegdam, R. Bonn and J. van der Elsken, *Chem. phys. lett.* **2**, 182 (1968).
24 C. Kittel, *Introduction to Solid State Physics*, John Wiley and Sons, New York (1966).
25 R. A. Levy, *Principles of Solid State Physics*, Academic Press, New York and London (1968).
26 M. Born and J. R. Oppenheimer, *Ann. Phys. Lpz.* **84**, 457 (1927).
27 H. C. Longuet-Higgins, in *Advances in Spectroscopy* vol. 2, p. 432, Wiley–Interscience, New York (1961).
28 J. N. Murrell, S. F. A. Kettle and J. M. Teddler, *Valence Theory*, John Wiley and Sons, London (1969).
29 J. M. Ziman, *Electrons and Phonons* chapter 1, Oxford University Press (1960).
30 A. A. Maradudin, E. W. Montroll and G. H. Weiss, *Theory of Lattice Dynamics in the Harmonic Approximation*, Solid state physics, supplement 3, Academic Press, New York (1963).
31 R. A. Smith, *Wave Mechanics of Crystalline Solids*, Chapman and Hall (1969).

References

32 B. D. Bartolo, *Optical Interactions in Solids* chapter 13, John Wiley and Sons, New York (1968).
33 H. Bilz, in reference 13, chapter 12.
34 M. Born and K. Huang, *Dynamical Theory of Crystal Lattices*, Oxford University Press (1954).
35 H. B. Rosenstock, *Phys. rev.* **121**, 416 (1961).
36 A. J. E. Foreman and W. M. Lomer, *Proc. phys. soc.* B **70**, 1143 (1957).
37 M. Born and T. von Kármán, *Physik. Z.* **13**, 297 (1912).
38 Reference 34, appendix 4.
39 J. Frenkel, *Wave Mechanics, Elementary Theory* p. 265, Oxford University Press (1932).
40 P. Debye, in reference 30, p. 9.
41 M. Born, in reference 30, p. 1.
42 W. Cochran, *Reports progr. phys.* **26**, 1 (1963).
43 Reference 30, all chapters.
44 G. Dolling and A. D. B. Woods, in *Thermal Neutron Scattering* chapter 5, edited by P. A. Egelstaff, Academic Press, New York (1965).
45 G. Dolling, in reference 16, p. 289.
46 R. P. Singh, *J. scient. ind. res.* **22**, 246 (1963).
47 G. Venkataraman and V. C. Sahni, *Rev. mod. Phys.* **42**, 409 (1970).
48 G. Leibfried and W. Ludwig, *Solid state phys.* **12**, 1 (1965).
49 R. A. Cowley, *Adv. in phys.* **12**, 421 (1963).
50 V. N. Kashcheev and M. A. Krivoglaz, *Sov. phys. solid state* **3**, 1107 (1961).
51 H. Bilz and L. Genzd, *Zeitschrift für Physik* **169**, 53 (1962).
52 R. P. Lowndes, *Phys. rev.* **1B**, 2754 (1970).
53 R. Nova, R. Callorotti, H. Ceva and A. Martinet, *Phys. rev.* **185**, 1177 (1969).
54 R. A. Cowley, *Rep. prog. phys.* **31**, 123 (1968).
55 A. S. Barker Jr, in *Far infra-red properties of solids* p. 297, edited by S. S. Mitra and S. Nudelman, Plenum Press, New York (1970).
56 T. Toya, *J. res. inst. catalysis, Hokkaido Univ.* **6**, 161 (1958).
57 T. Toya, *J. res. inst. catalysis, Hokkaido Univ.* **6**, 183 (1958).
58 T. Toya, *J. res. inst. catalysis, Hokkaido Univ.* **7**, 60 (1959).
59 T. Toya, in reference, 30, p. 91.
60 T. Toya, *Inelastic Scattering of Neutrons* vol. 1, p. 25, International Atomic Energy Agency Vienna (1965).
61 L. Sham, Ph.D. Thesis, Cambridge University (1963).
62 L. Sham, *Proc. Roy. Soc.* A **283**, 33 (1965).
63 W. A. Harrison, *Phys. rev.* **129**, 2503 (1963).
64 W. A. Harrison, *Phys. rev.* **129**, 2512 (1963).
65 W. A. Harrison, *Phys. rev.* **131**, 2433 (1963).
66 W. A. Harrison, *Phys. rev.* **136**, A 1107 (1964).
67 W. A. Harrison, *Phys. rev.* **139**, A 179 (1965).
68 W. A. Harrison, in reference 2, p. 73.
69 W. A. Harrison, in *Pseudopotentials in the Theory of Metals*, Benjamin, New York (1966).
70 S. H. Vosko, R. Taylor and G. H. Keech, *Can. J. phys.* **43**, 1187 (1965).
71 A. O. E. Animalu, F. Bonsignori and V. Bortolani, *Nuovo cimento* **42**B, 83 (1966).
72 A. O. E. Animalu, *Proc. Roy. Soc.* A **294**, 376 (1966).
73 R. Pick, *J. phys. (Paris)*, **28**, 539 (1967).
74 J. V. Koppel and A. A. Maradudin, *Phys. lett.* **24**A, 244 (1967).

References

75 V. C. Sahni, G. Venkataraman and A. P. Roy, *Phys. lett.* **23**, 633 (1966).
76 R. W. Shaw Jr and R. Pynn, *J. phys.* C **2**, 2071 (1969).
77 M. L. Cohen and V. Heine, *Solid state phys.* **24**, 37 (1970).
78 R. M. Pick, *Adv. phys.* **19**, 269 (1970).
79 F. A. Johnson, *Proc. Roy. Soc.* A **310**, 79, 89, 101 (1969).
80 P. D. De Cicco and F. A. Johnson, *Proc. Roy. Soc.* A **310**, 111 (1969).
81 W. Jones and N. H. March, *Proc. Roy. Soc.* A **317**, 359 (1970).
82 D. Pines, *Elementary excitations in solids*, Benjamin, New York (1963).
83 F. A. Johnson, *Proc. Roy. Soc.* A **317**, 279 (1970).
84 E. W. Kellermann, *Phil. trans. Roy. Soc.* A **238**, 513 (1940).
85 W. Cochran, *Phil. mag.* **4**, 1082 (1959).
86 W. Cochran, *Proc. Roy. Soc.* A **253**, 260 (1959).
87 A. D. B. Woods, W. Cochran and B. N. Brockhouse, *Phys. rev.* **119**, 980 (1960).
88 U. Schröder, *Solid state commun.* **4**, 347 (1966).
89 W. Cochran, in reference 30, p. 75.
90 W. Cochran, in reference 2, p. 53.
91 A. M. Karo and J. R. Hardy, *Phys. rev.* **141**, 696 (1966).
92 B. G. Dick and A. W. Overhauser, *Phys. rev.* **112**, 90 (1958).
93 K. P. Tolpygo and V. S. Mashkevich, *Sov. phys. JETP* **5**, 435 (1957).
94 J. R. Hardy, *Phil. mag.* **4**, 1278 (1959).
95 J. R. Hardy and A. M. Karo, *Phil. mag.* **5**, 859 (1960).
96 R. P. Lowndes and D. H. Martin, *Proc. Roy. Soc.* A **308**, 473 (1969).
97 B. G. Dick, in reference 30, p. 159.
98 G. Raunio and S. Rolandson, *Phys. rev.* **2**B, 2098 (1970).
99 V. S. Mashkevich and K. B. Tolpygo, *Sov. phys. JETP* **5**, 435 (1957).
100 K. B. Tolpygo, *Sov. phys. solid state* **3**, 685 (1962).
101 Z. A. Demidenko and K. B. Tolpygo, *Sov. phys. solid state* **3**, 2493 (1962).
102 V. S. Mashkevich, *Sov. phys. solid state* **2**, 2345 (1961).
103 H. Kaplan, in reference 30, p. 615.
104 S. K. Sinha, *Phys. rev.* **177**, 1256 (1969).
105 J. C. Phillips, *Phys. rev.* **166**, 832 (1968).
106 J. C. Phillips, *Phys. rev.* **168**, 917 (1968).
107 J. C. Phillips, *Phys. rev.* **168**, 904 (1968).
108 L. J. Sham, *Phys. rev.* **188**, 1431 (1969).
109 R. M. Pick, M. H. Cohen and R. M. Martin, *Phys. rev.* **1**B, 910 (1970).
110 S. K. Joshi and A. K. Rajagopal, *Solid state phys.* **22**, 159 (1968).
111 S. N. Singh and S. K. Joshi, *Physica* **47**, 277 (1970).
112 S. N. Singh and S. Prakash, *Physica* **50**, 10 (1970).
113 S. Prakash and S. K. Joshi, *Phys. rev.* **187**, 808 (1969).
114 I. G. Lang and U. S. Pashabekova, *Sov. Phys. solid state* **6**, 2913 (1965).
115 Ø. Ra, *J. chem. phys.* **52**, 3765 (1970).
116 T. Ōsaka, *J. chem. phys.* **54**, 863 (1971).
117 M. A. Nusinovici, M. Balkanski and J. L. Birman, *Phys. rev.* **1**B, 595 (1970).
118 O. Schnepp, in *Advances in Atomic and Molecular Physics* vol. 5, p. 155, Academic Press, New York and London (1969).
118a G. E. Leroid, *Trans. Am. cryst. soc.* **6**, 35 (1970).
119 D. N. Batchelder, M. F. Collins, B. C. G. Haywood and G. R. Sidey, *J. phys.* C **3**, 249 (1970).
120 G. L. Morley and K. L. Kliewer, *Phys. rev.* **180**, 245 (1969).
121 S. H. Walmsley, *Mol. phys.* **14**, 165 (1968).

References

122 T. S. Kuan, A. Warshel and O. Schnepp, *J. chem. phys.* **52**, 3012 (1970).
122a J. C. Lauffer and G. E. Leroi, *J. chem. phys.* **55**, 993 (1971).
123 M. C. A. Donkerslott and S. H. Walmsley, *Mol. phys.* **19**, 183 (1970).
124 O. Schnepp and A. Ron, *Diss. Faraday soc.* **48**, 26 (1969).
125 P. H. Martin and S. H. Walmsley, *Diss. Faraday soc.* **48**, 49 (1969).
126 G. S. Pawley, *Diss. Faraday soc.* **48**, 125 (1969).
127 S. H. Walmsley, in *Excitons, Magnons and Phonons in Molecular Crystals* p. 83, edited by A. B. Zahlan, Cambridge University Press (1968).
128 M. V. Bobetic and J. A. Barker, *Phys. rev.* **1**B, 4169 (1970); M. L. Klein, J. A. Barker and T. R. Koehler, *Phys. rev.* **4**B, 1983 (1971).
129 J. D. Axe, G. Sherane and K. A. Müller, *Phys. rev.* **183**, 820 (1969).
130 J. B. Bates, E. R. Lippincott, Y. Mikawa and R. J. Jakobsen, *J. chem. phys.* **52**, 3731 (1970).
131 N. E. Tornberg and C. H. Perry, *J. chem. phys.* **53**, 2946 (1970).
132 E. W. Small, B. Fanconi and W. L. Peticolas, *J. chem. phys.* **52**, 4369 (1970).
133 J. M. Dickey and A. Paskin, *Phys. rev.* **188**, 1407 (1969).
134 K. N. Klump, O. Schnepp and L. H. Nosanow, *Phys. rev.* **1**B, 2496 (1970).
135 M. L. Klein and T. R. Kochler, *J. phys.* C **3**, L102 (1970).
136 T. Nakamura, *Progr. theoret. phys.* (*Kyoto*) **14**, 135 (1955).
137 Lord Rayleigh, *Proc. London math. soc.* **17**, 4 (1885).
138 A. E. H. Love, *Some problems of geodynamics* p. 160, Cambridge University Press (1911).
139 R. E. Lee and R. M. White, *Appl. phys. lett.* **12**, 12 (1968).
140 D. C. Auth, *Appl. phys. lett.* **16**, 521 (1970).
141 G. Cachier, *Appl. phys. lett.* **17**, 419 (1970).
142 F. E. Goodwin and M. E. Pendinoff, *Appl. phys. lett.* **8**, 69 (1966).
143 M. B. Schulz, B. J. Matsinger and M. G. Holland, *J. appl. phys.* **41**, 2755 (1970)
144 C. C. Tseng, *J. appl. phys.* **41**, 2270 (1970).
145 R. E. Allen, G. P. Alldredge and F. W. de Wette, *Phys. rev. lett.* **23**, 1285 (1969).
146 R. E. Allen, G. P. Alldredge and F. W. de Wette, *Phys. rev. lett.* **24**, 301 (1970).
147 Reference 30, chapter 6.
148 D. C. Gazis, R. Herman and R. F. Wallis, *Phys. rev.* **119**, 533 (1960).
149 D. C. Gazis and R. F. Wallis, *J. math. phys.* **3**, 190 (1962); R. E. Allen, G. P. Alldredge and F. W. de Wette, *Phys. rev.* **4**B, 1661 (1971).
150 A. A. Lucas, *J. chem. phys.* **48**, 3156 (1968).
151 S. Y. Tong and A. A. Maradudin, *Phys. rev.* **181**, 1318 (1969).
152 B. N. N. Achar and G. R. Barsch, *Phys. rev.* **188**, 1361 (1969).
153 L. Dobrzynski and D. L. Mills, *J. phys. chem. solids* **30**, 1043 (1969); G. P. Alldredge, R. E. Allen, and F. W. de Wette, *Phys. rev.* **4**B, 1682 (1971).
154 B. Lengelar and W. Ludwig, *Phys. stat. solidi* **7**, 463 (1964).
155 A. A. Maradudin, *Solid state phys.* **19**, 1 (1966).
156 C. H. Hodges, *Phys. rev.* **187**, 994 (1969).
157 J. W. Christian and V. Vítek, *Rep. prog. phys.* **33**, 307 (1970).
158 D. W. Pashley, *Rep. prog. phys.* **28**, 291 (1965).
159 S. Amelinckx, in *The interaction of radiation with solids* p. 682, North-Holland Publishing Company, Amsterdam (1964).
160 P. Dean, in *Lattice Dynamics* p. 561, edited by R. F. Wallis, Pergamon Press, London (1964).
161 A. A. Maradudin, *Rep. prog. phys.* **28**, (1965).

References

162 I. M. Lifshitz and A. M. Kosevich, *Rep. prog. phys.* **29**, 217 (1966).
163 A. A. Maradudin, *Solid state phys.* **18**, 273 (1966).
164 L. Genzel, in reference 12, chapter 15.
165 R. J. Elliott, in reference 2, p. 377.
166 C. W. McCombie, in reference 55, p. 297.
167 K. K. Rebane, *Impurity Spectra of Solids*, translated from Russian by J. S. Shier, Plenum Publishing Corporation, New York (1970).
168 R. C. Newman, *Adv. phys.* **18**, 545 (1969).
169 P. Dean and M. D. Bacon, *Proc. Roy. Soc.* A **283**, 64 (1965).
170 E. W. Montroll and R. B. Potts, *Phys. rev.* **100**, 525 (1955).
171 P. Mazur, E. W. Montroll and R. B. Potts, *J. Wash. acad. sci.* **46**, 2 (1956).
172 R. L. Bjork, *Phys. rev.* **105**, 456 (1957).
173 M. Wagner, *Phys. rev.* **131**, 2520 (1963).
174 B. Szigeti, *Phys. chem. solids* **24**, 225 (1963).
175 K. Dettmann and W. Ludwig, *Phys. stat. solidi* **10**, 689 (1965); R. S. Singh and S. S. Mitra, *Phys. rev.* **5B**, 733 (1972).
176 R. W. Munn, *J. chem. phys.* **52**, 64 (1970).
177 M. Wagner, *Phys. rev.* **133**, 750 (1964).
178 M. V. Klein, *Phys. rev.* **141**, 716 (1966).
179 O. Litzman and P. Rozsa, *Proc. phys. Soc.* **85**, 285 (1965).
180 K. Kunc, *Czech. J. phys.* B **15**, 883 (1965).
181 J. A. Krumhanzl and J. A. D. Matthew, *Phys. rev.* **166**, 856 (1968).
182 J. Mahanty, *Phys. lett.* **19**A, 583 (1969).
183 M. Sachclev and J. Mahanty, *J. phys.* C **3**, 1225 (1970).
184 J. H. Weiner and W. F. Adler, *Phys. rev.* **144**, 511 (1966).
185 D. H. Tsai, R. Bullough and R. C. Perrin, *J. phys.* C**3**, 2022 (1970); B. K. Agrawal, *Phys rev.* **3**B, 1843 (1971).
186 R. Brout and W. Visscher, *Phys. rev. lett.* **9**, 54 (1962).
187 Yu. M. Kagan and Ya. A. Iosilevskii, *Sov. phys. JETP.* **18**, 162 (1965).
188 S. Takeno, *Progr. theoret. phys. (Kyoto)* **29**, 191 (1963).
189 J. G. Kirkwood, *J. chem. phys.* **7**, 506 (1939).
190 K. S. Pitzer, *J. chem. phys.* **8**, 711 (1940).
191 S. Krimm, C. Y. Liang and G. B. B. M. Sutherland, *J. chem. phys.* **25**, 549 (1956).
192 C. Tric, *J. chem. phys.* **51**, 4778 (1969), and **55**, 827 (1971).
193 J. H. Schachtschneider, *Shell report*, 57/65.
194 J. H. Schachtschneider and R. G. Snyder, *Spectrochim. acta*, **19**, 117 (1963).
195 R. G. Snyder and J. H. Schachtschneider, *Spectrochim. acta*, **20**, 853 (1964).
196 L. Pizieri and G. Zerbi, *J. chem. phys.* **48**, 3561 (1968).
196a I. R. Beattie, N. Cheetham, T. R. Gilson, K. M. S. Livingston and D. J. Reynolds, *J. chem. soc.* (*A*) 1910 (1971).
197 R. Tubino and G. Zerbi, *J. chem. phys.* **51**, 4509 (1969).
198 P. M. A. Sherwood and J. J. Turner, *Spectrochim. acta*, **26**A, 1975 (1970).
199 A. Axman, W. Biem, P. Borsch, F. Hoßfeld and H. Steller, *Diss. Faraday soc.* **48**, 69 (1969).
200 M. Tasumi and T. Shimanouchi, *J. chem. phys.* **43**, 1245 (1965).
201 R. Zbinden, *Infra-red spectroscopy of high polymers*, Academic Press, New York and London (1964).
202 G. Zerbi, in *Advances in Applied Spectroscopy* vol. 2, edited by E. G. Brame, Marcel Dekker, New York (1969).

References

203 G. Zerbi, in reference 55, p. 497.
204 D. O. Hummel, *Infrared Spectra of Polymers*, Wiley–Interscience, London (1966).
205 A. Elliott, *Infra-red spectra and structure of organic long-chain polymers*, E. Arnold, London (1969).
206 G. Zerbi, *Appl. spect. rev.* **2**, 193 (1969).
207 D. A. Long, in *Ann. reports* A **65**, 83 (1968).
208 A. Anderson, H. A. Gebbie and S. H. Walmsley, *Mol. phys.* **7**, 401 (1964).
209 E. A. Moelwyn-Hughes, *Physical Chemistry*, Pergamon Press, London (1961).
210 J. N. Plendl, in reference 12, p. 309.
211 J. N. Plendl, *Phys. rev.* **119**, 1598 (1960).
212 J. N. Plendl and P. J. Gielisse, *Appl. opt.* **4**, 853 (1965).
213 S. P. Srivastava and M. P. Madan, *J. phys. soc. Japan* **23**, 459 (1967).
214 S. P. Srivastava and M. P. Madan, *J. phys. soc. Japan* **23**, 1433 (1967).
215 R. J. Seeger and E. Teller, *Phys. rev.* **62**, 37 (1942).
216 E. W. Nuffield, *X-ray diffraction methods* chapter 8, John Wiley and Sons, New York (1966).
217 J. E. Bertie and R. Kopelman, *J. chem. phys.* **55**, 3613 (1971).
218 B. N. Brockhouse, S. Hautecler and H. Stiller, in reference 159, p. 580.
219 J. W. White, in reference 127, p. 43.
220 J. W. White, in reference 16, p. 463.
221 P. A. Egelstaff, in reference 16, p. 135.
222 G. F. Longster and J. W. White, *Mol. phys.* **17**, 1 (1969).
223 K. A. Strong, R. M. Brugger and R. J. Pugmire, *J. chem. phys.* **52**, 2277 (1970).
224 K. W. Logan, S. F. Trevino, H. J. Prask and J. D. Gault, *J. chem. phys.* **53**, 3417 (1970).
225 M. F. Collins and B. C. Haywood, *J. chem. phys.* **52**, 5740 (1970); N. A. Lurie and H. R. Danner, *J. chem. phys.*, **55**, 4156 (1971).
226 A. J. Maeland, *J. chem. phys.* **52**, 3952 (1970).
226a A. J. Maeland and D. E. Holmes, *J. chem. phys.* **54**, 3979 (1971).
226b B. L. Deopura and V. D. Gupta, *J. chem. phys.* **54**, 4013 (1971).
227 G. C. Stirling, C. J. Ludman and T. C. Waddington, *J. chem. phys.* **52**, 2730 (1970).
227a U. Efron, I. Pelah, U. Vulkan and H. Zafrir, *J. chem. phys.* **55**, 3599 (1971).
227b D. H. Day and R. N. Sinclair, *J. chem. phys.* **55**, 2807 (1971).
227c J. N. Plendl, P. J. Gielisse, H. S. Plendl, R. A. Kromhout and L. C. Mansur, *Appl. opt.* **10**, 1444 (1971).
228 F. A. Cotton, *Chemical Applications of Group Theory*, Wiley–Interscience, New York (1963).
229 D. Schonland, *Molecular Symmetry*, Van Nostrand, London (1965).
230 D. F. Johnston, *Rep. prog. phys.* **23**, 66 (1960).
231 H. Jones, in reference 16, p. 1.
232 V. Heine, *Group theory in quantum mechanics – an introduction to its present usage*, International monographs in pure and applied mathematics, Pergamon Press, London (1960).
233 M. Tinkham, *Group theory and quantum mechanics*, International series in pure an applied physics, McGraw-Hill, New York (1964).
234 E. B. Wilson, J. C. Decius and P. C. Cross, *Molecular Vibrations – The theory of infra-red and Raman vibrational spectra*, McGraw-Hill, London (1955).
235 S. Bhagavantam and T. Venkatarayudu, *Theory of Groups and its application to physical problems*, Academic Press, New York and London (1969).

References

236 H. Winston and R. S. Halford, *J. chem. phys.* **17**, 607 (1948).
237 R. S. Halford, *J. chem. phys.* **14**, 8 (1946).
238 G. K. Koster, *Solid state phys.* **5**, 174 (1957).
239 M. V. Bobetic and J. A. Barker, *Phys. rev.* **2B**, 4169 (1970).
240 C. J. Bradley and A. P. Cracknell, *The Mathematical Theory of Symmetry in Solids*, Oxford University Press (1970).
241 S. Bhagavantam, *Crystal symmetry and physical problems*, Academic Press, London and New York (1966).
242 R. Kopelman, *J. chem. phys.* **47**, 2631 (1967).
243 J. E. Bertie and J. W. Bell, *J. chem. phys.* **54**, 160 (1971).
244 E. R. Bernstein, S. D. Colson, R. Kopelman and G. W. Robinson, *J. chem. phys.* **48**, 5596 (1968).
245 *International Tables for X-ray Crystallography* vol. 1, Kynoch Press, Birmingham (1952).
246 D. M. Adams and D. C. Newton, *Tables for Factor Group Analysis*, Beckman—RIIC, London (1970).
247 D. M. Adams and Newton, *J. chem. Soc. (A)*, 2822 (1970).
248 C. J. Bradley and A. P. Cracknell, *J. phys.* C **3**, 610 (1970).
249 S. A. Pollack, *J. chem. phys.* **38**, 98 (1963).
250 S. Bhagavantam and T. Venkatarayudu, *Proc. Ind. acad. sci.* **9A**, 224 (1939).
251 M. Balkanski and M. K. Teng, in *Physics of the Solid State*, p. 289, Commemoration Volume to Professor S. Bhagavantam, edited by S. Balakrishna, M. Krishnamurthi and R. R. Rao, Academic Press, London (1969).
252 Reference 228, p. 257.
253 Reference 32, p. 234.
254 G. Herzberg, *Molecular spectra and molecular structure, II, Infrared and Raman spectra of polyatomic molecules* p. 179, Van Nostrand, Princeton, New Jersey (1945).
255 W. L. Peticolas, L. Nafie, P. Stein and B. Fanconi, *J. chem. phys.* **52**, 1576 (1970).
256 L. I. Schiff, *Quantum Mechanics*, 2nd edition, McGraw-Hill, New York (1955).
257 Reference 256, p. 398.
258 Reference 254, p. 125
259 J. L. Birman, *Phys. rev.* **127**, 1093 (1962).
260 J. L. Birman, *Phys. rev.* **131**, 1489 (1963).
261 J. L. Birman, *Phys. rev.* **150**, 771 (1966).
262 J. L. Birman, M. Lax and R. Loudon, *Phys. rev.* **145**, 620 (1966).
263 M. A. Nusimovici and J. L. Birman, *Phys. rev.* **156**, 925 (1967).
264 E. Burstein, F. A. Johnson and R. Loudon, *Phys. rev.* **139A**, 1239 (1965).
265 S. Ganesan and E. Burstein, *J. phys. (Paris)* **26**, 645 (1965).
266 F. A. Johnson and R. Loudon, *Proc. Roy. Soc.* A **231**, 274 (1964).
267 S. S. Mitra, *Phys. lett.* **11**, 119 (1964).
268 O. Oehler and H. H. Günthard, *J. chem. phys.* **51**, 4714 (1969).
269 K. E. Lawson, *Infra-red absorption of inorganic substances*, Reinhold Publishing Corporation, New York (1961).
270 The Chemical Society – Recent Specialist Periodical Reports, *Spectroscopic properties of inorganic and organometallic compounds*; J. J. Turner, *Chem. Ind*, 109, (1966).
271 H. W. Schrötter, in *Raman spectroscopy, theory and practice* vol. 2, p. 69, edited by H. A. Szymanski, Plenum Press, New York (1970).
272 R. E. Hester, in reference 271.
273 W. P. Griffith, *J. chem. soc. (A)* 286 (1970).

References

274 J. R. Ferraro, in reference 55, p. 475.
275 D. Bloor, *Infrared phys.* **10**, 1 (1970).
276 K. Ohwada, *Spectrochim. acta* **26**A, 1035 (1970).
277 K. Ohwada, *Spectrochim. acta* **26**A, 1723 (1970).
278 R. Forneris, J. Hiraishi, F. A. Miller and M. Uzhora, *Spectrochim. acta* **26**A, 581 (1970).
279 F. Bessette, A. Cabana, R. P. Rournier and R. Savoie, *Canad, J. chem.* **48**, 410 (1970).
280 S. A. Miller, H. E. Rast and H. H. Caspers *J. chem. phys.* **52**, 4172 (1970).
281 J. F. Scott, *J. chem. phys.* **53**, 852 (1970).
282 R. Summitt, *Bull. Am. phys. soc.* **11**, 829 (1966).
283 Z. Iqbal, C. W. Brown and S. S. Mitra, *J. chem. phys.* **52**, 4867 (1970); Z. Iqbal and M. L. Malhotra, *J. chem. phys.* **55**, 528 (1971).
284 J. B. Bates, A. S. Quist and G. E. Boyd, *J. chem. phys.* **54**, 124 (1971).
285 S. A. Miller, H. H. Caspers and H. E. Rast, *Phys. rev.* **168**, 964 (1968).
286 H. H. Caspers, R. A. Buchanan and H. R. Merlin, *J. chem. phys.* **41**, 94 (1964).
287 R. P. Lowndes, J. F. Parrish and C. H. Perry, *Phys. rev.* **182**, 913 (1969).
288 D. M. Adams and R. R. Smardzewski, *J. chem. Soc.* (A) 8 (1971).
289 G. A. Ozin, *Canad. J. chem.* **48**, 2931 (1970).
290 J. Maillols, L. Bardet and R. Marignan, *J. chim. phys.* **66**, 529 (1969).
291 N. T. McDevitt, *J. opt. soc. Amer.* **59**, 1240 (1970).
292 J. F. McCaffrey, N. T. McDevitt and C. M. Phillippi, *J. opt. soc. Amer.* **61**, 209 (1971).
293 O. Oehler and H. H. Günthard, *J. chem. phys.* **51**, 4714 (1969).
294 J. L. Verbel and R. F. Wallis, *Phys. rev.* **182**, 783 (1969).
295 M. H. Brooker, D. E. Irish and G. E. Boyd, *J. chem. phys.* **53**, 1083 (1970).
296 K. M. M. Kruse, *Spectrochim. acta*, **26**A, 1603 (1970).
297 A. Hadni, in reference 127, p. 31.
298 J. E. Cahill and G. E. Leroi, *J. chem. phys.* **51**, 1324 (1969).
299 J. E. Cahill and G. E. Leroi, *J. chem. phys.* **51**, 97 (1969).
300 A. Anderson and L. Y. Wong, *Canad. J. chem.* **47**, 2713 (1969).
301 G. Lucovsky, *Bull. Am. phys. soc.* **12**, 102 (1967).
302 G. Lucovsky and R. C. Keezer, *Solid state commun.* **5**, 439 (1967).
303 C. H. Wong and P. A. Gleury, *J. chem. phys.* **53**, 2243 (1970).
304 R. Savoie and M. Pérzolet, *J. chim. phys.* **50**, 2781 (1969).
305 H. B. Friedrich and R. E. Carlson, *J. chem. phys.* **53**, 4441 (1970).
306 P. Taimsalu and D. W. Robinson, *Spectrochim. acta*, **21**, 1921 (1965).
307 A. Anderson and S. H. Walmsley, *Mol. phys.* **10**, 391 (1966).
308 A. J. Melveger, J. W. Brasch and E. R. Lippincott, *Appl. opt.* **9**, 11 (1970).
309 J. E. Cahill and G. E. Leroi, *J. chem. phys.* **51**, 4514 (1969).
310 V. Wagner, *Phys. lett.* **22**, 58 (1966).
311 J. L. Hustan and H. C. Claasen, *J. chem. phys.* **52**, 5646 (1970).
312 R. D. McLachlan and V. B. Carter, *Spectrochim. acta*, **26**A, 1121 (1970).
313 M. P. Marzocchi and P. Manzelli, *J. chem. phys.* **52**, 2630 (1970).
314 D. M. Thomas, J. B. Bates, A. Bandy and E. R. Lippincott, *J. chem. phys.* **53**, 3698 (1970).
315 H. F. Shurvell and J. A. Faniran, *J. mol. spec.* **33**, 436 (1970).
316 M. Ito, T. Yokoyama and M. Susuki, *Spectrochim. acta*, **26**A, 695 (1970).
317 Y. Mikawa and R. J. Jakobsen, *J. mol. spec.* **33**, 178 (1970).
318 C. Di Lauro, S. Califano and G. Adembri, *J. mol. struc.* **2**, 173 (1968).

References

319 Y. Mikawa, J. W. Brasch and R. J. Jakobsen, *J. mol. struc.* **3**, 103 (1969).
320 R. D. McLachlan and V. B. Carter, *Spectrochim. acta*, **26**A, 2247 (1970).
321 C. W. Brown, R. J. Obermski and E. R. Lippincott, *J. chem. phys.* **52**, 2253 (1970).
322 D. A. Oliver and S. H. Walmsley, *Mol. phys.* **17**, 617 (1969).
323 I. Harada and T. Shimanouchi, *J. chem. phys.* **55**, 3605 (1971).
324 J. Loisel and V. Lorenzelli, *J. mol. struc.* **1**, 157 (1967).
325 R. Claus, H. W. Schrötter, J. Brandmüller and S. Haüsshl, *J. chem. phys.* **52**, 6448 (1970).
326 A. Brce and R. A. Kydel, *Spectrochim. acta*, **26**A, 1791 (1970).
327 F. G. Baglin and C. B. Roze, *Spectrochim. acta*, **26**A, 2293 (1970).
328 B. Lunelli and C. Pecile, *J. chem. phys.* **52**, 2375 (1970); T. Takenaka, *Spectrochim. acta*, **27**A, 1735, (1971).
329 D. M. Adams and A. Squire, *J. chem. soc.* (A) 814 (1970).
330 E. Gazis, P. Heim, C. Meister and F. Dorr, *Spectrochim. acta*, **26**A, 497 (1970).
331 E. W. Small, B. Fanconi and W. L. Peticolas, *J. chem. phys.* **52**, 4369 (1970).
332 V. Schettino, M. P. Marzocchi and G. Sbrana, *J. mol. struc.* **2**, 39 (1968).
333 I. Harada and R. C. Lord, *Spectrochim. acta*, **26**A, 2305 (1970); H. Susi and J. S. Ard, *Spectrochim. acta*, **27**A, 1549 (1971).
334 N. Sheppard, *Trans. Faraday soc.* **51**, 9 (1965).
335 P. B. Jamieson, S. C. Abrahams and J. C. Bernstein, *J. chem. phys.* **48**, 5058 (1969) and **50**, 4352 (1969).
336 S. D. Ross, *J. phys.* C **3**, 1785 (1970).
337 G. Natta, *Gazz. chim. Ital.* **60**, 911 (1930).
338 M. Atoji and W. N. Lipscomb, *Acta. cryst.* **7**, 597 (1954).
339 D. E. Sands, A. Zalkin and R. E. Elson, *Acta, cryst.* **12**, 21 (1969).
340 I. R. Beattie, K. M. S. Livingston, D. J. Reynolds and G. A. Ozin, *J. chem. soc.* (A) 1210 (1970).
341 B. I. Bleaney and B. Bleaney, *Electricity and Magnetism* p. 241, Oxford University Press (1957).
342 K. Huang, *Proc. Roy. Soc.* A **208**, 352 (1951).
343 R. H. Lyddane, R. G. Sachs and E. Teller, *Phys. rev.* **59**, 673 (1941).
344 B. Szigeti, *Proc. phys. soc.* A **204**, 51 (1951).
345 S. P. Srivastava and M. P. Madan, *J. phys. soc. Japan*, **23**, 459 (1967).
346 W. Cochran, *Z. Kristallogr.* **112**, 30 (1959).
347 W. Cochran, *Adv. phys.* **9**, 387 (1960).
348 L. Merton, *Z. Naturf.* A **15**, 47 (1960).
349 W. Cochran and R. A. Cowley, *J. phys. chem. solids*, **23**, 447 (1962).
350 Reference 341, p. 492.
350a H. Barentzen, *J. chem. phys.* **55**, 3664 (1971).
351 B. Szigetti, *Trans. Faraday soc.* **45**, 155 (1949).
352 R. F. Guertin and F. Stern, *Phys. rev.* **134**, A 427 (1964).
353 O. Theimer and R. Paul, *J. appl. phys.* **36**, 3678 (1965).
354 C. Haas and J. A. A. Ketelaar, *Physica*, **22**, 1286 (1956).
355 E. J. Ambrose, A. Elliot and R. B. Temple, *Proc. Roy. Soc.* A **206**, 192 (1951).
356 G. C. Pimentel and A. L. McClellan, *J. chem. phys.* **20**, 270 (1952).
357 C. Haas and D. F. Hornig, *J. chem. phys.* **26**, 707 (1957).
357a A. Frech and J. C. Decius, *J. chem. phys.* **54**, 2374 (1971).
358 J. J. Hopfield, *Phys. rev.* **112**, 1555 (1958).

References

359 U. Fano, *Phys. rev.* **103**, 1202 (1956).
360 J. N. Plendl, in reference 12, p. 387.
361 J. N. Plendl, S. S. Mitra and P. J. Gielisse, *Phys. stat. solidi*, **12**, 367 (1965).
362 J. N. Plendl, *Phys. rev.* **123**, 1172 (1961).
363 A. A. Maradudin and S. Ushioda, *J. phys. chem. solids*, **31**, 1075 (1970).
364 C. W. Cleek, F. J. P. Consitt and W. D. Lawson, *Infrared phys.* **5**, 141 (1965).
365 I. Simon, *J. opt. soc. Amer.* **41**, 336 (1951).
366 T. S. Robinson and W. C. Price, *Proc. phys. soc.* B **66**, 969 (1953).
367 P. N. Schatz, S. Maeda and K. Kozima, *J. chem. phys.* **38**, 2658 (1963).
368 A. A. Maradudin and R. F. Wallis, *Phys. rev.* **125**, 4 (1962).
369 V. V. Mitskevich, *Sov. phys. solid state*, **4**, 2224 (1963).
370 O. J. Neuberger and R. D. Hatcher, *J. chem. phys.* **34**, 5 (1961).
371 R. Wehner, *Phys. stat. solidi*, **15**, 725 (1966).
372 R. A. Cowley, *Advances in physics*, **12**, 421 (1963).
372a A. A. Maradudin and R. F. Wallis, *Phys. rev.* **2**B, 4294 (1970); R. F. Wallis and A. A. Maradudin, *Phys. rev.* **3**B, 2063 (1971).
373 R. Loudon, *Proc. Roy. Soc.* A **275**, 218 (1963).
374 J. L. Birman and A. K. Ganguly, *Phys. rev. lett.* **17**, 647 (1966).
375 A. K. Ganguly and J. L. Birman, *Phys. rev.* **162**, 806 (1967).
376 L. N. Ovander, *Sov. phys. solid state*, **3**, 1737 (1962).
377 L. N. Ovander, *Sov. phys. solid state*, **4**, 1081 (1962).
378 L. N. Ovander, *Sov. phys. solid state*, **8**, 1939 (1967).
379 E. Burnstein, D. L. Mills, A. Pinczuk and S. Ushioda, *Phys. rev. lett.* **22**, 348 (1969).
380 D. L. Mills and E. Burnstein, *Phys. rev.* **188**, 1465 (1969).
380a B. Bendow, *Phys. rev.* **2**B, 5051 (1970).
381 A. S. Pine and G. Dresselhaus, *Phys. rev.* **188**, 1489 (1969).
382 L. Couture-Mathieu, H. Poulet and J. P. Mathieu, *Comp. rend.* **234**, 1761 (1952).
383 O. Brafman and S. S. Mitra, *Phys. rev.* **171**, 931 (1968).
384 J. C. Irwin and J. LaCombe, *Canad. J. phys.* **48**, 2499 (1970).
385 J. C. Irwin and J. LaCombe, *J. appl. phys.* **41**, 1444 (1970).
386 C. H. Henry and J. J. Hopfield, *Phys. rev. lett.* **15**, 964 (1965).
387 S. P. S. Porto, B. Tell and T. C. Daman, *Phys. rev.* **16**, 450 (1966).
388 J. F. Scott, L. E. Cheesman and S. P. S. Porto, *Phys. rev.* **162**, 834 (1967).
389 H. E. Puthoff, R. H. Pantell, H. E. Puthoff and J. M. Yarborough, *Appl. phys. lett.* **14**, 258 (1969).
390 S. K. Kurtz and J. A. Giordmaine, *Phys. rev. lett.* **22**, 192 (1969).
391 R. Claus, H. W. Schrötter, H. H. Hacker and S. Haussühl, *Z. Naturf.* A **24**, 1733 (1969).
392 R. Claus, *Z. Naturf.* A **25**, 306 (1970).
393 L. Merten, *Z. Naturf.* A **24**, 1878 (1969).
394 R. Claus, *Phys. lett.* **31**A, 299 (1970).
395 D. A. Long, *Chemistry in Britain*, **7**, 108 (1971).
396 J. Gelbwachs, R. H. Pantell, H. E. Puthoff and J. M. Yarborough, *Appl. phys. lett.* **14**, 258 (1969).
397 J. M. Yarborough, S. S. Sussman, H. E. Puthoff, R. H. Pantell and B. C. Johnson, *Appl. phys. lett.* **15**, 102 (1969).
398 M. K. Srivastava. *J. opt. soc. amer.* **60**, 1542 (1970).
399 C. M. Hartwig, D. L. Rousseau and S. P. S. Porto, *Phys. rev.* **188**, 1328 (1969).

References

400 C. A. Arguello, D. L. Rousseau and S. P. S. Porto, *Phys. rev.* **181**, 1351 (1969).
401 L. Merten, *Phys. stat. solidi*, **28**, 111 (1968).
402 D. J. Olechna, *J. phys. chem. solids*, **31**, 2755 (1970).
403 T. C. Damen, S. P. S. Porto and B. Tell, *Phys. rev.* **142**, 570 (1966).
404 R. K. Khanna and P. J. Miller, *Spectrochim. acta*, **26**A, 1667 (1970).
405 L. Couture-Mathieu, J. A. A. Ketelaar, W. Vedder and J. Fahrenfort, *J. chem. phys.* **20**, 1492 (1952).
406 J. F. Scott and S. P. S. Porto, *Phys. rev.* **1**B, 2818 (1970).
407 L. Merten, *Z. Naturf.* A **23**, 1183 (1968).
408 J. Onstott and G. Lucovsky, *J. phys. chem. solids*, **31**, 2171 (1970).
409 W. Otaguro, C. A. Arguello and S. P. S. Porto, *Phys. rev.* **1**B, 2818 (1970).
409a M. Ishigame, T. Sato and T. Sakurai, *Phys. rev.* **3**B, 4388 (1971).
410 L. Merten, *Z. Naturf.* A **22**, 359 (1967).
411 C. K. Asawa, *Phys. rev.* **2**B, 2068 (1970).
412 G. L. Bottger and A. L. Geddes, *J. chem. phys.* **46**, 3000 (1967).
413 T. H. K. Barron, *Phys. rev.* **123**, 1995 (1961).
414 R. Ruppin and R. Englman, *Rep. prog. phys.* **33**, 149 (1970).
415 R. Englman and R. Ruppin, *J. phys.* C **1**, 614 (1968).
416 R. Fuchs and K. L. Kliewer, *Phys. rev.* **140** A 2076 (1965).
417 D. W. Berreman, *Phys. rev.* **130**, 2193 (1963).
418 H. Fröhlich, *Theory of Dielectrics*, Oxford University Press (1949).
419 I. Richman, *J. chem. phys.* **41**, 2836 (1964).
420 H. D. Riccius, *J. appl. phys.* **39**, 4381 (1968).
421 F. Priox and M. Balkanski, *Phys. stat. solidi*, **32**, 119 (1969).
422 K. L. Kliewer and R. Fuchs, *Phys. rev.* **144**, 495 (1966).
423 K. L. Kliewer and R. Fuchs, *Phys. rev.* **150**, 573 (1966).
424 R. Fuchs and K. L. Kliewer, *J. opt. soc. Amer.* **58**, 319 (1968).
425 R. Englman and R. Ruppin, *Localized Excitations in Solids* p. 41, edited by R. F. Wallis, Plenum Press, New York (1967).
426 R. Ruppin and R. Englman, *J. phys.* C **1**, 630 (1968).
427 R. Fuchs, K. L. Kliewer and W. J. Pardee, *Phys. rev.* **150**, 589 (1966).
428 D. Fox and R. M. Hexter, *J. chem. phys.* **41**, 1125 (1964).
429 R. K. Bullough and A. S. F. Obada, *Chem. phys. lett.* **3**, 114 (1969).
430 R. K. Bullough and A. S. F. Obada, *Chem. phys. lett.* **3**, 177 (1969).
431 R. K. Bullough and B. V. Thompson, *J. phys.* C **3**, 1780 (1970).
432 M. Tsuyobi, M. Terada and T. Shimanouchi, *J. chem. phys.* **36**, 1301 (1962).
433 T. P. Martin, *Phys. rev.* **177**, 1349 (1969).
434 G. R. Hunt, C. H. Perry and J. Ferguson, *Phys. rev.* **134**, A 688 (1969).
435 A. J. Slobodnik Jr, *Appl. phys. lett.* **14**, 94 (1969).
436 A. J. Slobodnik Jr, P. H. Carr and A. J. Budreau, *J. appl. phys.* **41**, 4380 (1970).
437 H. Boersch, J. Geiger and W. Stickel, *Phys. rev. lett.* **17**, 379 (1966).
438 H. Boersch, J. Greiger and W. Stickel, *Z. Phys.* **212**, 130 (1968).
439 H. Ibach, *Phys. rev. lett.* **24**, 1416 (1970).
439a J. B. Chase and K. L. Kliewer, *Phys. rev.* **2**B, 4389 (1970); A. A. Lucus and M. Sunjic, *Phys. rev. lett.* **26**, 229 (1971).
440 Reference 29, chapter 3.
441 F. Seitz, *The Modern Theory of Solids*, McGraw-Hill, London and New York (1940).
442 D. A. Kleinman, *Phys. rev.* **118**, 118 (1960).
443 J. R. Jasperse, A. Kahan, J. N. Plendl and S. S. Mitra, *Phys. rev.* **146**, 526 (1966).

References

444 J. N. Plendl, *Appl. optics*, **10**, 87 (1971).
445 A. A. Maradudin, *Phys. stat. solidi*, **2**, 1493 (1962).
446 A. A. Maradudin and A. E. Fein, *Phys. rev.* **128**, 2589 (1962).
447 I. P. Ipatova, A. A. Maradudin and R. F. Wallis, *Phys. rev.* **155**, 882 (1967).
448 R. A. Cowley, *J. phys. (Paris)* **26**, 659 (1965).
449 G. Dolling and R. A. Cowley, *Proc. phys. soc.* **88**, 463 (1966).
450 R. F. Wallis, I. P. Ipatova and A. Maradudin, *Sov. phys. solid state*, **8**, 850 (1966).
451 D. W. Jepsen and R. F. Wallis, *Phys. rev.* **125**, 1496 (1962).
452 M. Hass, *Phys. rev.* **117**, 1497 (1960).
453 L. E. Gurevich and I. P. Ipatova, *Sov. physics JETP*, **18**, 162 (1965).
454 K. V. Krishna Rao, in reference 251, p. 415.
455 E. L. Slaggie, *Phys. rev.* **2**B, 2230 (1970).
456 F. Freund, *Spectrochim. acta*, **26**A, 195 (1970).
457 R. V. St. Louis and O. Schnepp, *J. chem. phys.* **50**, 5177 (1969).
458 Y. A. Schwartz, A. Ron and S. Kimel, *J. chem. phys.* **54**, 99 (1971).
459 J. F. Scott, *Phys. rev. lett.* **21**, 907 (1968).
460 G. Burns and B. A. Scott, *Phys. rev. lett.* **25**, 167 (1970).
461 J. F. Scott and J. P. Remeika, *Phys. rev.* **1**B, 4182 (1970).
462 G. P. O'Leary and R. G. Wheeler, *Phys. rev.* **1**B, 4409 (1970); D. K. Agrawal and C. H. Perry, *Phys. rev.* **4**B, 1893 (1971).
462a J. C. Raich and R. D. Etters, *J. chem. phys.* **55**, 3901 (1971).
463 G. O. Jones, D. H. Martin, P. A. Mawer and C. H. Perry, *Proc. Roy. Soc.* A **261**, 10 (1961).
464 M. Lax and E. Burstein, *Phys. rev.* **97**, 39 (1955).
465 J. F. Scott, R. C. C. Leite and T. C. Damen, *Phys. rev.* **188**, 1285 (1969).
466 Reference 29, chapter 3.
467 W. L. Peticolas, *Ann. rev. phys. chem.* **18**, 233 (1967).
468 M. H. Cohen and J. Ruvalds, *Phys. rev. lett.* **23**, 1378 (1969).
469 S. S. Mitra, *J. chem. phys.* **39**, 3031 (1963).
470 L. Van Hove, *Phys. rev.* **89**, 1189 (1953).
471 J. C. Phillips, *Phys. rev.* **104**, 1263 (1956).
472 R. Loudon and F. A. Johnson, *Proc. Roy. Soc.* A **281**, 274 (1964).
473 P. Denham, G. R. Field, P. L. R. Morse and G. R. Wilkinson, *Proc. Roy. Soc* A **317**, 55 (1970).
474 J. C. Phillips, *Phys. rev.* **113**, 149 (1959).
475 H. E. Rast, H. H. Caspers and S. A. Miller, *Phys. rev.* **180**, 894 (1969).
476 A. Nedoluha, *Phys. rev.* **1**B, 864 (1970).
476a N. B. Manson, W. van der Ohe and S. L. Chodos, *Phys. rev.* **3**B, 1968 (1971).
477 M. J. Stephen, *Proc. phys. soc.* **71**, 485 (1958).
478 L. Kleinmann, *Bull. Amer. phys. soc.* **10**, 392 (1965).
479 Reference 24, p. 166.
480 J. Rovalds and A. Zawadowski, *Phys. rev.* **2**B, 1172 (1970).
481 D. L. Rousseau and S. P. S. Porto, *Phys. rev. lett.* **20**, 1354 (1968).
482 J. E. Bertie and E. Whalley, *J. chem. phys.* **46**, 1271 (1967).
483 N. T. Melamed, *J. appl. phys.* **34**, 560 (1963).
484 C. M. Phillippi, in *Developments in Applied Spectroscopy* vol. 7 B, Pergamon Press, New York (1970).
484a S. J. Allen, Jr and H. J. Guggenheim, *Phys. rev.* **4**B, 937 and 950 (1971).
485 P. M. A. Sherwood, *Spectrochim. acta*, **27**A, 1019 (1971).

References

486 G. E. Ewing and G. C. Pimentel, *J. chem. phys.* **35**, 925 (1961).
487 H. E. Hallam, in *Spectroscopy* p. 245, edited by M. J. Wells, Institute of Petroleum, London (1962).
488 A. J. Barnes and H. E. Hallam, *Quart. rev.* **23**, 392 (1969).
489 C. A. Stevenson, W. B. Person, D. A. Dows and R. M. Hexter, *J. chem. phys.* **31**, 1324 (1959).
490 J. K. Burdett, *J. mol. spec.* **36**, 365 (1970).
491 D. F. Hornig, *Disc. Faraday soc.* **9**, 115 (1950).
492 I. I. Mador and R. S. Quinn, *J. chem. phys.* **20**, 1837 (1952).
493 R. M. Hexter, *J. chem. phys.* **25**, 1286 (1956).
494 S. Zwerdling and R. S. Halford, *J. chem. phys.* **23**, 2221 (1955).
495 J. R. Durig, D. J. Antion and C. B. Pate, *J. chem. phys.* **51**, 4449 (1969).
496 D. A. Dows, *J. chem. phys.* **29**, 484 (1959).
496a F. J. Boerio and J. L. Koenig, *J. chem. phys.* **52**, 3425 (1970), and **54**, 3667 (1971).
497 D. R. Hornig and W. W. Osberg, *J. chem. phys.* **23**, 662 (1955).
498 W. Vedder, *The infrared absorption spectrum of the ammonium ion in lattices of the NaCl type*, Strudentendrukkerij Poortpers N.V. (1958).
499 W. Vedder and D. F. Hornig, *J. chem. phys.* **35**, 1560 (1961).
499a K. B. Harvey and N. R. McQuaker, *J. chem. phys.* **55**, 4390 (1971).
500 J. Preudhomme and P. Tarte, *Spectrochim. acta.* **27**A, 845 (1971).
501 P. Tarte and J. Preudhomme, *Spectrochim. acta.* **26**A, 2207 (1970).
502 N. Sheppard, *J. mol. struc.* **6**, 5 (1970).
503 D. M. Grant, R. J. Pugmire, R. C. Livingston, K. A. Strong, H. L. McMurry and R. M. Brugger, *J. chem. phys.* **52**, 4424 (1970).
504 J. R. Durig, S. M. Craven and J. Bragin, *J. chem. phys.* **52**, 2046 (1970).
505 G. S. Pawley and S. J. Cyvin, *J. chem. phys.* **52**, 4073 (1970).
506 A. Anderson, H. A. Gebbie and S. H. Walmsley, *Mol. phys.* **7**, 401 (1964).
507 E. Sandor, *Acta cryst.* **15**, 463 (1962).
508 R. A. Schroider, C. E. Weir and E. R. Lippincott, *J. res. nat. bur. stand.* **66**A, 407 (1962).
509 R. M. Hexter and D. A. Dows, *J. chem. phys.* **25**, 504 (1956).
510 N. E. Schumaker and C. W. Gorland, *J. chem. phys.* **53**, 392 (1970).
511 J. R. Durig, C. B. Pate and Y. S. Li, *J. chem. phys.* **54**, 1033 (1971).
512 J. Van Kranendank, *Canad. J. phys.* **38**, 240 (1960).
513 H. M. James, *Phys. rev.* **2**B, 2213 (1970).
514 J. C. Raich and H. M. James, *Phys. rev. lett.* **16**, 173 (1966).
515 J. C. Raich and R. D. Etters, *Phys. rev.* **168**, 425 (1968).
516 C. F. Coll III, A. B. Harris and A. J. Berlinsky, *Phys. rev. lett.* **25**, 858 (1970).
517 C. F. Coll III and A. B. Harris, *Phys. rev.* **2**B, 1176 (1970); A. J. Berlinsky and A. B. Harris, *Phys. rev.* **4**B, 2808 (1971).
518 M. Clouter and H. P. Gush, *Phys. rev. lett.* **15**, 200 (1965).
519 W. N. Hardy, I. F. Silvera and J. P. McTague, *Phys. rev. lett.* **22**, 297 (1969), and **26**, 127 (1971).
520 E. J. Allin, T. Feldman and H. L. Welsh, *J. chem. phys.* **24**, 1116 (1956).
521 J. van Kranendonk and G. Karl, *Rev. mod. phys.* **40**, 531 (1968).
522 J. Noolandi, *Canad. J. phys.* **48**, 2032 (1970).
523 H. P. Gush, W. F. J. Hare, E. J. Allin and H. L. Welsh, *Canad. J. phys.* **38**, 176 (1960).
524 H. L. Welsh, E. J. Allin and V. Soots, in reference 251, p. 343.
525 H. R. Zaidi, *Can. J. phys.* **48**, 1539 (1970).

References

526 J. T. Luxton, D. J. Montgomery and R. Summitt, *Phys. rev.* **188**, 1345 (1969).
527 P. Coufová and J. Novák, *Infrared phys.* **8**, 153 (1968).
528 Reference 234, p. 162.
529 H. Yamada and W. B. Person, *J. chem. phys.* **41**, 2478 (1964).
530 P. N. Schatz, *J. chem. phys.* **32**, 894 (1960).
531 P. N. Schatz, *J. chem. phys.* **31**, 1146 (1959).
532 E. Klein, J. P. le Geroz, O. R. Trautz and R. Z. le Geroz, in reference 484.
533 J. Overend, in *Infra-red spectroscopy and molecular structure* p. 345, edited by M. Davies, Elsevier (1963).
534 J. Burdett, *Chem. phys. lett.* **5**, 10 (1970).
535 K. G. Brown and W. T. King, *J. chem. phys.* **52**, 4437 (1970).
536 J. W. Smith, *Electric dipole moments*, Butterworths, London (1955).
537 P. R. Davies and W. J. Orville-Thomas, *J. mol. struc.* **4**, 163 (1969); G. A. Thomas, G. Jalsovszky, J. A. Ladd and W. J. Orville-Thomas, *J. mol. struc.* **8**, 1 (1971).
538 A. R. Bandy, H. B. Friedrich and W. B. Person, *J. chem. phys.* **53**, 674 (1970).
539 A. D. Buckingham, *Proc. Roy. Soc.* A **248**, 169 (1958).
540 A. D. Buckingham, *Proc. Roy. Soc.* A **255**, 32 (1960).
541 H. B. Friedrich, *J. chem. phys.* **52**, 3005 (1970).
542 H. B. Friedrich and R. E. Carlson, *J. chem. phys.* **53**, 4441 (1970).
543 F. W. DeWette and G. E. Schacher, *Phys. rev.* **137**A 78 (1965).
544 M. Mandel and P. Mazur, *Physica*, **24**, 116 (1958).
545 O. Schnepp, *J. chem. phys.* **46**, 3983 (1967).
546 M. Ito, M. Suzuki and T. Yokoyama, *J. chem. phys.* **50**, 249 (1969).
547 H. B. Friedrich, *J. chem. phys.* **47**, 4269 (1967).
548 J. G. David, and W. G. Person, *J. chem. phys.* **48**, 510 (1968).
548a R. E. Carlson and H. B. Friedrich, *J. chem. phys.* **54**, 2794 (1971).
549 R. V. St Louis and O. Schnepp, *J. chem. phys.* **50**, 5177 (1969).
550 M. Britt and A. Ron, *J. chem. phys.* **50**, 3053 (1969).
551 K. G. Brown and W. T. King, *J. chem. phys.* **52**, 4437 (1970).
552 L. A. Woodward, in *Raman Spectroscopy*, p. 1 edited by H. A. Szymanski, Plenum Press, New York (1967).
553 G. Placzek, in *Handbuch der Radiologie* vol. 6, edited by E. Marx, Akademische Verlagsgesellschaft, Leipzig (1934).
554 R. E. Hester, in reference 552, p. 101.
555 H. M. J. Smith, *Phil. trans.* A **241**, 105 (1948).
556 Reference 24, p. 199.
557 R. A. Cowley, *Proc. phys. soc.* **84**, 281 (1964).
558 W. D. Johnston Jr, *Phys. rev.* **1**B, 3494 (1970).
559 W. F. Murphy, W. Holzer and H. J. Bernstein, *Appl. Spectry.* **23**, 211 (1969).
560 J. Tang and A. C. Albrecht, in reference 271.
561 J. E. Cahill and G. E. Leroi, *J. chem. phys.* **51**, 4514 (1969).
562 J. Brandmüller and R. Claus, *Spectrochim. acta*, **25**A, 103 (1969).
563 M. V. Klein and S. P. S. Porto, *Phys. rev. lett.* **22**, 782 (1969).
564 M. P. Fontana and E. Mulazzi, *Phys. rev. lett.* **25**, 1102 (1970).
565 J. F. Scott, *Phys. rev.* **2**B, 1209 (1970).
566 D. C. Hamilton, *Phys. rev.* **188**, 1221 (1969).
567 E. Mulazci, *Phys. rev. lett.* **25**, 228 (1970).
568 R. M. Martin and T. C. Damen, *Phys. rev. lett.* **26**, 86 (1971).
568a B. Bendow, *Phys. rev.* **4**B, 552 (1971); B. Bendow and J. L. Birman, *Phys. rev.* **4**B, 569 (1971).

References

569 J. Behringer, in reference 552, p. 168.
570 S. M. Shapiro, R. W. Gammon and H. Z. Cummins, *Appl. phys. lett.* **9**, 157 (1966).
571 G. Durand and A. S. Pine, *J. quantum electronics*, **4**, 525 (1968).
572 A. S. Pine, *Phys. rev.* **185**, 1187 (1969).
573 A. S. Pine, in *Light Scattering Spectra of Solids*, edited by G. B. Wright, Springer Verlag, New York (1969).
574 D. Pohl and W. Kaiser, *Phys. rev.* **1B**, 31 (1970).
575 P. Flubacher, A. J. Leadbetter, J. A. Morrison and B. P. Stoicheff, *J. phys. chem. solids*, **12**, 53 (1959).
576 J. Comly, E. Gormire and A. Yariv, *J. appl. phys.* **38**, 4091 (1967).
577 F. Harrigan, C. Klein, R. Rudko and D. Wilson, *Microwaves*, **8**, 68 (1969).
578 J. E. Kiefer and A. Yariv, *Appl. phys. lett.* **15**, 26 (1969).
579 R. Weil, *J. appl. phys.* **40**, 2857 (1969).
580 W. B. Gandrud, *Appl. opt.* **9**, 1936 (1970).
581 J. N. Plendl, *Appl. opt.* **9**, 2768 (1970).
582 J. Fahrenfort, *Spectrochim. acta*, **17**, 698 (1961).
583 I. Newton, *Opticks* (1717), Dover Publications, New York (1952).
584 N. J. Harrick, *Internal Reflection Spectroscopy*, Wiley–Interscience, New York (1967).
585 P. A. Wilks Jr and T. Hirschfeld, *Appl. spectry. rev.* **1**, 99 (1967).
586 N. J. Harrick and N. H. Riederman, *Spectrochim. acta*, **21**, 2135 (1965).
587 P. A. Flournoy and W. J. Schaffers, *Spectrochim. acta*, **22**, 5 (1966).
588 J. P. Devlin, G. Pollard and R. French, *J. chem. phys.* **53**, 4147 (1970).
589 K. Tsuji, H. Yamada, K. Suzuki and I. Nitta, *Spectrochim. acta* **26A**, 475 (1970).
590 R. R. Joyce and P. L. Richards, *Phys. rev. lett.* **24**, 1007 (1970).
591 A. B. Pippard, *The Dynamics of Conduction Electrons* p. 64, Gordon and Bread, New York (1965).
592 T. Holstein, *Phys. rev.* **88**, 1425 (1952).
593 R. B. Dingle, *Physica*, **19**, 311 (1953).
594 T. Holstein, *Phys. rev.* **96**, 535 (1954).
595 H. Scher, *Phys. rev. lett.* **25**, 759 (1979); P. B. Allen, *Phys. rev.* **3B**, 305 (1971); H. Scher, *Phys. rev.* **3B**, 3551 (1971).
596 *How to select infra-red crystals*, Barnes Engineering Company, Stamford, Connecticut 06902; D. E. McCarthy, *Appl. opt.* **10**, 2539 (1971).
597 A. Hadni, in reference 55, p. 561.
598 S. P. S. Porto and J. F. Scott, *Phys. rev.* **157**, 716 (1967).
598a F. Cerdeira, C. J. Buchenauer, F. H. Pollak and M. Cordona, *Phys. rev.* **5B**, 580 (1972).
599 J. F. Scott, *J. chem. phys.* **48**, 874 (1968).
600 A. S. Barker Jr, *Phys. rev.* **132**, 1474 (1964).
601 J. R. Ferraro, S. S. Mitra and C. Postmus, *Inorg. nucl. chem. lett.* **2**, 269 (1966) and **4**, 55 (1968).
602 L. J. Basile, C. Postmus and J. R. Ferraro, *Spectry. lett.* **1**, 189 (1968).
603 J. R. Ferraro, C. Postmus and S. S. Mitra, *Phys. rev.* **174**, 983 (1968).
604 J. R. Ferraro, S. S. Mitra and C. Postmus, *Phys. rev. lett.* **18**, 455 (1967).
605 J. R. Ferraro, C. Postmus, S. S. Mitra and C. J. Hoskins, *Appl. opt.* **9**, 5 (1970).
606 T. H. K. Barron and A. Batana, *Phil. mag.* **20**, 619 (1969).
607 J. R. Ferraro, S. S. Mitra, C. Postmus, C. J. Hoskins and E. C. Siwiec, *Appl. spectry.* **24**, 187 (1970).

References

608 S. S. Mitra, O. Brafman, W. B. Daniels and R. K. Crawford, *Phys. rev.* **186**, 942 (1969).
609 J. R. Ferraro, *J. chem. phys.* **53**, 117 (1970).
610 R. Eckhardt, D. Eggers and L. J. Slutsky, *Spectrochim. acta*, **26**A, 2033 (1970).
611 J. R. Ferraro, in reference 55, p. 451.
612 W. B. Daniels, in reference 30, p. 273.
613 D. H. Saunderson, *Phys. rev. lett.* **17**, 530 (1966).
614 R. Lechner and G. Quittner, *Phys. rev. lett.* **17**, 1259 (1966).
615 J. A. Leake, W. B. Daniels, J. Skalyo Jr, B. C. Frazer and G. Shirane, *Phys. rev.* **181**, 1251 (1969).
616 R. T. Payne, *Phys. rev. lett.* **13**, 53 (1964).
617 C. C. Bradley, *High pressure methods in solid state research*, Plenum Press, New York (1969).
618 C. C. Bradley, H. A. Gebbie, A. C. Gilby, V. V. Kechin and J. H. King, *Nature*, **211**, 839 (1966).
619 R. J. Jakobsen and Y. Mikawa, *Appl. opt.* **9**, 17 (1970).
620 E. Anastassakis, S. Iwasa and E. Burstein, *Phys. rev. lett.* **17**, 1051 (1966).
621 P. A. Fleury and J. M. Worlock, *Phys. rev. lett.* **18**, 665 (1967).
622 J. M. Worlock and P. A. Fleury, *Phys. rev. lett.* **19**, 1176 (1967).
623 R. F. Schaufele, M. J. Weber and B. D. Silverman, *Phys. rev. lett.* **19**, 47 (1967).
624 G. P. Vella-Coleiro, *Phys. rev. lett.* **23**, 697 (1969).
624a J. Smit, *Phys. rev.* 3B, 4330 (1971).
625 D. A. Long, *Chemistry in Britain*, **7**, 108 (1971).
626 D. A. Long, *Essays in structural chemistry* chapter 2, Macmillan, London (1971).
627 S. J. Cyvin, J. E. Rauch and J. C. Decius, *J. chem. phys.* **43**, 4083 (1965).
628 J. H. Christie and D. L. Lockwood, *J. chem. phys.* **54**, 1141 (1971).
629 S. F. Fischer, G. L. Hofacker and M. A. Ratner, *J. chem. phys.* **52**, 1934 (1970).
630 C. Haas and D. F. Hornig, *J. chem. phys.* **32**, 1763 (1960).
631 D. Hadzi, *Hydrogen Bonding*, Pergamon Press, New York (1959).
632 G. C. Pimentel and A. L. McClellan, *The Hydrogen Bond*, Freeman, San Francisco (1960).
633 W. C. Hamilton and J. A. Ibers, *Hydrogen Bonding in Solids*, Benjamin, New York (1967).
634 T. R. Singh and J. L. Wood, *J. chem. phys.* **48**, 4567 (1968).
635 Y. Marechal and Q. Witkowski, *Theoret. chim. acta*, **9**, 116 (1967).
636 L. J. Bellamy and A. J. Owen, *Spectrochim. acta*, **25**A, 329 (1969).
637 C. N. R. Rao and A. S. N. Murphy, in reference 484.
638 A. S. N. Murphy and C. N. R. Rao, *Appl. Specty. rev.* **2**, 69 (1969).
638a G. R. Anderson and E. R. Lippincott, *J. chem. phys.* **55**, 4077 (1971).
639 E. Whalley and J. E. Bertie, *J. chem. phys.* **46**, 1264 (1967); E. Whalley and H. J. Labbé, *J. chem. phys.* **51**, 3120 (1969); D. D. King and E. Whalley, *J. chem. phys.* **56**, 553 (1972).
640 J. E. Bertie and E. Whalley, *J. chem. phys.* **46**, 1271 (1967); J. E. Bertie, H. J. Labbé and E. Whalley, *J. chem. phys.* **50**, 4501 (1969).
641 I. G. Austin and E. S. Gorbett, *Phil. mag.* **23**, 17 (1971).
642 M. Tasumi and G. Zerbi, *J. chem. phys.* **48**, 3813 (1970).
642a V. M. Bermudez, *J. chem. phys.* **54**, 4150 (1971).
643 A. Hadni, J. Claudel, X. Gerbaux, G. Marlet and J. M. Munier, *Appl. opt.* **4**, 487 (1965).
644 M. C. Tobin and T. Back, *J. opt. soc. Amer.* **58**, 1459 (1968).

References

645 R. J. Bell, N. F. Bind and P. Dean, *Proc. phys. soc.* C **1**, 299 (1968).
646 M. Hass, *J. phys. chem solids*, **31**, 415 (1970).
647 M. C. Tobin and T. Bååk, *J. opt. soc. Amer.* **60**, 368 (1970).
648 R. Shuker and R. W. Gammon, *Phys. rev. lett.* **25**, 222 (1970).
649 *Physics of Colour Centres*, edited by W. B. Fowler, Academic Press, New York (1968).
650 P. G. Dawber and R. J. Elliott, *Proc. phys. soc.* **81**, 453 (1963).
651 R. O. Davies and D. Healey, *J. phys.* C **1**, 1184 (1968); P. D. Mannheim and II. Friedmann, *Phys. stat. solidi*, **39**, 409, 1970; P. D. Mannheim, *Phys. rev.* **5B**, 745 (1972).
652 J. F. Angress and S. D. Smith, *Phys. lett.* **6**, 131 (1963).
653 J. F. Angress, A. R. Goodwin and S. D. Smith, *Proc. Roy. Soc.* A **287**, 64 (1965).
654 R. S. Leigh and B. Szigeti, *J. phys.* C **3**, 782 (1970).
655 G. O. Jones and J. M. Woodfine, *Proc. phys. soc.* **86**, 101 (1965).
656 J. Obriot, P. Marteau, H. Vu and B. Bodar, *Spectrochim. acta*, **26**A, 2051 (1970).
657 J. de Remigis and H. L. Welsh, *Canad. J. phys.* **48**, 1622 (1970).
658 J. S. Ogden and J. J. Turner, *Chemistry in Britain*, 186 (1971).
659 H. Friedmann and S. Kimel, *J. chem. phys.* **41**, 2552 (1964), and **43**, 3952 (1965), and **44**, 4359 (1966).
660 D. W. Robinson and W. G. von Holle, *J. chem. phys.* **44**, 410 (1966); M. G. Mason, W. G. von Holle and D. W. Robinson, *J. chem. phys.* **54**, 3491 (1971).
661 K. Pandey and S. Chandra, *J. chem. phys.* **45**, 4369 (1966).
662 R. V. St Louis and B. Crawford, *J. chem. phys.* **42**, 857 (1965).
663 M. A. Cundill and W. F. Sherman, *Phys. rev.* **168**, 1007 (1968).
664 M. V. Klein, in reference 649, p. 430; C. G. Olson and D. W. Lynch, *Phys. rev.* **4B**, 1990 (1971).
665 J. B. Page Jr and B. G. Dick, *Phys. rev.* **163**, 910 (1967).
666 M. E. Striefler and S. S. Jaswal, *Phys. rev.* **185**, 1194 (1969).
667 G. Benedek and G. F. Nardelli, *Phys. rev.* **154**, 872 (1967).
668 G. Benedek and E. Mulazzi, *Phys. rev.* **179**, 906 (1969).
669 R. W. Alexander Jr, A. E. Hughes and A. J. Sievers, *Phys. rev.* **1B**, 1563 (1970).
670 A. E. Hughes, *Contemporary Phys.* **12**, 231 (1971).
671 M. V. Klein, R. Wedding and M. A. Levine, *Phys. rev.* **180**, 902 (1969); G. K. Pandey and D. K. Shukle, *Phys. rev.* **3B**, 4391 (1971)
672 E. H. Coker and D. E. Hofer, *J. chem. phys.* **48**, 2713 (1968).
673 T. Mauring, *Eesti NSV Tead, Akad. Toim.* **17**, 203 (1968).
674 M. S. Pidzirailo and I. M. Khalimonova, *Ukrain, fiz. Zhur.* **12**, 1063 (1967).
675 R. Metselaar and J. van der Elsken, *Phys. rev.* **165**, 359 (1968).
676 W. A. Morgan, E. Silberman and H. W. Morgan, *Spectrochim. acta*, **23**A, 2855 (1967).
677 A. I. Stekhanov and M. B. Eliashberg, *Sov. phys. solid state*, **6**, 11 (1965).
678 A. I. Stekhanov and T. I. Maksimova, *Sov. phys. solid state*, **8**, 737 (1967).
679 L. C. Kravitz, *Phys. rev. lett.* **24**, 884 (1970); R. T. Harley, J. B. Page Jr and C. T. Walker, *Phys. rev.* **3B**, 1365 (1971).
680 W. Holzer, W. F. Murphy, H. J. Bernstein and J. Rolfe, *J. mol. spectry.* **26**, 534 (1968).
681 J. Rolfe, W. Holzer, W. F. Murphy and H. J. Bernstein, *J. chem. phys.* **49**, 936 (1968).
682 W. Holzer, W. F. Murphy and H. J. Bernstein, *J. mol. spectry.* **32**, 13 (1969).
683 I. F. Chang and S. S. Mitra, *Phys. rev.* **172**, 924 (1968).

References

684 J. H. Fertel and C. H. Perry, *Phys. rev.* **184**, 874 (1969).
685 I. F. Chang and S. S. Mitra, *Phys. rev.* **2B**, 1215 (1970).
686 J. J. Markham, *F-centres in Alkali Halides*, Solid State Physics Supplement, Academic Press, New York (1966).
687 D. Baüerle and B. Fritz, *Solid state commun*, **6**, 453 (1968); D. Baüerle and R. Hübner, *Phys. rev.* **2B**, 4252 (1970).
688 R. S. Singh and S. S. Mitra, *Phys. rev.* **2B**, 1070 (1970).
689 J. M. Worlock and S. P. S. Porto, *Phys. rev. lett.* **15**, 697 (1965).
690 C. J. Buchenaur, D. B. Fitchen and J. B. Page Jr, in reference 573.
691 P. Eisenberger and M. G. Adlerstein, *Phys. rev.* **1B**, 1787 (1970).
692 D. L. Mills and A. A. Maradudin, *Phys. rev.* **1B**, 903 (1970).
692a I. T. Jacobs, G. D. Jones, K. Zdansky and R. A. Satten, *Phys. rev.* **3B**, 2888 (1971).
693 C-K. Chau and M. V. Klein, *Phys. rev.* **1B**, 2642 (1970).
694 J. A. Harrington and C. T. Walker, *Phys. rev.* **1B**, 903 (1970).
695 B. K. Agrawal, *J. phys.* C **3**, 1002 (1970).
696 D. L. Dexter and R. S. Knox, *Excitons*, Wiley–Interscience, New York (1965).
697 D. P. Craig and S. H. Walmsley, *Excitons in Molecular Crystals*, Benjamin, New York (1968).
698 R. S. Knox, *The Theory of Excitons*, Solid State Physics Supplement no. 5, Academic Press, New York (1963).
699 S. Nikitine, in reference 12.
700 D. C. Reynolds, in reference 12.
701 S. Nikitine, *Memoires de la Societe Royale des sciences de Liege*, **20**, 201 (1970).
702 A. S. Davydov, *Theory of Molecular Excitons*, McGraw-Hill, New York (1962).
703 G. Baldine and B. Bosacchi, *Memoires de la Societe Royale des sciences de Liege*, **20**, 305 (1970).
704 D. S. McClure, in reference 12.
705 R. R. Joyce and P. L. Richards, *Phys. rev.* **179**, 375 (1969).
706 G. A. Slack, S. Roberts, and F. S. Ham, *Phys. rev.* **155**, 170 (1967); J. T. Vallin, G. A. Slack and C. C. Bradley, *Phys. rev.* **2B**, 4406 (1970).
707 X. Gerbaux and A. Hadni, *J. chim. phys.* **67**, 1674 (1970).
708 A. Hadni, in reference 55, p. 535.
709 J. A. Koningstein, in reference 55, p. 516.
710 J. A. Koningstein and Toaning-Ng, *J. opt. soc. Amer.* **58**, 1462 (1968).
711 P. Grunberg and J. A. Koningstein, *J. chem. phys.* **53**, 4584 (1970).
712 G. Schaack and J. A. Koningstein, *J. opt. soc. Amer.* **60**, 1110 (1970).
713 Y-N. Chiu, *J. chem. phys.* **52**, 3641 (1970).
714 P. L. Richards and M. Tinkham, *Phys. rev.* **119**, 575 (1960).
715 A. A. Abrikosov and L. A. Falkovskii, *Sov. phys. JETP*, **13**, 179 (1961).
716 Y. Toyozawa, *Prog. theor. phys. (Kyoto)*, **12**, 421 (1953).
717 *Polarons and Excitons*, edited by C. G. Kuper and G. D. Whitfield, Oliver and Boyd, Edinburgh (1963).
718 J. Appel, *Solid state phys.* **21**, 193 (1968).
719 I. G. Austin and N. F. Mott. *Adv. in phys.* **18**, 41 (1969).
720 J. T. Devreese, L. F. Lemmens, R. Evrard and E. Kartheuser, *Memoires de la Societe Royale des sciences de Liege*, **20**, 249 (1970).
721 R. Mühlstroh and H. G. Reik, *Phys. rev.* **162**, 703 (1967).
722 P. Gerthsen, R. Groth, K. H. Härdtl, D. Heese and H. G. Reik, *Solid state commun.* **3**, 165 (1965).

References

723 H. G. Reik, E. Kauer and P. Gerthsen, *Phys. lett.* **8**, 29 (1964).
724 H. G. Reik and R. Mühlstroh, *Solid state commun.* **5**, 105 (1967).
725 A. S. Barker Jr, in *Optical Properties and Electronic Structure of Metals and Alloys*, p. 160, North-Holland, Amsterdam (1966).
726 V. N. Bogomolov, E. K. Kodinov, D. N. Mirlin and Y. A. Firsov, *Sov. phys. solid state*, **9**, 2077 (1967).
727 E. K. Kudinov, D. N. Mirlin and Y. A. Firsov, *Sov. phys. solid state*, **11**, 2257 (1970).
728 I. G. Austin, B. D. Clay, C. E. Turner and A. J. Springthrope, *Solid state commun.* **6**, 53 (1968).
729 N. S. Hush, in *Progress in Inorganic Chemistry* p. 357, edited by F. A. Cotton, vol. 8, John Wiley and Sons, London (1967).
730 P. Eisenberger, P. S. Pershan and D. R. Bosernworth, *Phys. rev.* **188**, 1197 (1969).
731 P. Eisenberger and M. G. Adlerstein, *Phys. rev.* **1B**, 1787 (1970).
732 H. Finkenrath, N. Uhle and W. Waidelich, *Solid state commun.* **7**, 11 (1969).
733 R. C. Brandt and F. C. Brown, *Phys. rev.* **181**, 1241 (1969).
734 R. C. Brandt, D. M. Larsen, P. P. Crooker and G. B. Wright, *Phys. rev. lett.* **23**, 240 (1969).
735 D. Bohm and D. Pines, *Phys. rev.* **82**, 625 (1951).
736 I. I. Sobelnon and E. L. Feinbing, *Sov. phys. JETP*, **13**, 179 (1961).
737 N. Tzoar and E-N. Foo, *Phys. rev.* **180**, 535 (1969).
738 E. N. Economou, *Phys. rev.* **182**, 539 (1969).
739 D. C. Tsui and A. S. Barker Jr, *Phys. rev.* **186**, 590 (1969).
740 S. N. Jasperson and S. E. Schnatterly, *Phys. rev.* **188**, 759, (1969).
741 J. L. Stanford, *J. opt. soc. Amer.* **60**, 49 (1970).
742 C. G. Olson and D. W. Lynch, *Phys. rev.* **177**, 1231 (1969).
743 A. Mooradian and G. B. Wright, *Phys. rev. lett.* **16**, 999 (1966).
744 A. Mooradian and A. L. McWhorter, *Phys. rev. lett.* **19**, 849 (1967); J. F. Scott, T. C. Damen, J. Ruvelds and A. Zawadowski, *Phys. rev.* **3B**, 1295, (1971).
745 R. R. Alfano, *J. opt. soc. Amer.* **60**, 66 (1970).
746 C. K. N. Patel and R. E. Slusher, *Phys. rev. lett.* **22**, 282 (1969); J. Shah, T. C. Damen, J. F. Scott and R. C. C. Leite, *Phys. rev.* **3B**, 4238 (1971).
747 M. Tinkham, in reference 55, p. 196.
748 A. J. Sievers, *J. appl. phys.* **41**, 980 (1970).
749 A. J. Sievers and M. Tinkham, *Phys. rev.* **124**, 321 (1961).
750 P. L. Richards, *J. appl. phys.* **34**, 1237 (1963).
751 R. C. Ohlmann and M. Tinkham, *Phys. rev.* **123**, 425 (1961).
752 P. L. Richards, *J. appl. phys.* **34**, 1237 (1963).
753 T. G. Blocker, M. A. Kinch and F. G. West, *Phys. rev. lett.* **22**, 853 (1969).
754 P. A. Fleury, S. P. S. Porto, L. E. Cheesman and H. J. Guggenheim, *Phys. rev. lett.* **17**, 84 (1966).
755 R. M. Macfarlane, *Phys. rev. lett.* **25**, 1454 (1970).
756 R. Loudon, *Adv. phys.* **17**, 243 (1968).
757 J. W. Halley and I. Silvera, *Phys. rev. lett.* **15**, 654 (1965).
758 J. W. Halley, *Phys. rev.* **149**, 423 (1966).
759 Y. Tanabe, T. Moriya and S. Sugano, *Phys. rev. lett.* **15**, 1023 (1965).
760 J. W. Halley, *Phys. rev.* **154**, 458 (1967).
761 S. J. Allen Jr, R. Loudon and P. L. Richards, *Phys. rev. lett.* **16**, 463 (1966).
762 P. A. Fleury, *Bull. Amer. phys. soc.* **12**, 420 (1967).

References

763 P. L. Richards, *Bull. Amer. phys. soc.* **10**, 33 (1965).
764 P. A. Fleury, S. P. S. Porto and R. Loudon, *Phys. rev. lett.* **18**, 658 (1967).
765 R. J. Elliott, M. F. Thorpe, G. F. Imbusch, R. Loudon and J. B. Parkinson, *Phys. rev. lett.* **21**, 147 (1968).
766 P. A. Fleury, *Phys. rev. lett.* **21**, 151 (1968).
767 A. Oseroff and P. S. Pershon, *Phys. rev. lett.* **21**, 1593 (1968).
768 P. Moch, G. Parisot, R. E. Dietz and H. J. Guggenheim, *Phys. rev. lett.* **21**, 1596 (1968).
769 M. F. Thorpe, *Phys. rev. lett.* **23**, 472 (1969).
770 P. A. Fleury, J. M. Worlock and H. J. Guggenheim, *Phys. rev.* **185**, 738 (1969).
771 J. B. Torrance Jr and M. Tinkham, *Phys. rev.* **187**, 587 (1969).
772 J. B. Torrance Jr and M. Tinkham, *J. appl. phys.* **39**, 822 (1968).
773 J. B. Torrance Jr and M. Tinkham, *Phys. rev.* **187**, 595 (1969).
774 M. Tinkham, in reference 55, p. 223.
775 D. L. Mills and S. Ushioda, *Phys. rev.* **2B**, 3805 (1970).
776 S. R. Chinn, *Phys. rev.* **3B**, 1 (1971).
777 S. R. Chinn, H. J. Zeiger and J. R. O'Connor, *Phys. rev.* **3B**, 1709 (1971).
778 R. E. Dietz, G. I. Parisot and A. E. Meixner, *Phys. rev.* **4B**, 2302 (1971).

Glossary of terms

Symbol	Explanation	Page where symbol is introduced
a	separation of atoms	8
$\mathbf{a}$	unit lattice vector of magnitude a	8
a'	half width of crystal slab	117
a''	length of unit cell side	132
a^+	creation operator	149
a^-	annihilation operator	149
A	amplitude	5
A	number of ions pairs per unit volume	87
$A(\nu)$	absorption	95
A'	absorption coefficient	103
A	anharmonic interaction	141
A''	modified absorption coefficient	167
A_i	number of ion pairs or molecules in a unit volume for an initial state	167
A_f	number of ion pairs or molecules in a unit volume for a final state	167
A^d	number of defects per unit volume	189
B	bulk mode	132
c	velocity of light	8
c'	longitudinal stiffness	8
C	virtual mode that is a solution to the cotangent equation	125
d_e	effective thickness	178
d_p	penetration depth	178
d	atomic orbital,	197
D	non-linear dielectric interaction	141
D	exciton and polaron term	196, 199
e	electronic charge	200
E	electric field	75
E_l	local field	91
E_1	field from atoms outside Lorentz sphere	91
E_2	field from atoms inside Lorentz sphere	91
E	electric mode	124
ΔE	splitting of degenerate states	132
E_r	field of incident radiation	165
E_i	macroscopic electric field inside particle	165
E_p	electric field due to plasmon	200
$f(\nu)$	reflectivity or absorption coefficient	40
F	factor group	48
F'	factor group of S'	53
$F(A)$	factor of anharmonicity	100
g	force constant	10

Glossary of terms

Symbol	Explanation	Page where symbol is introduced
$g(\nu)$	density of states	38
g	size parameter	132
G	reciprocal lattice vector	44
$\left.\begin{array}{l}h\\ \hbar\end{array}\right\}$	Planck's constant	25
h_j	number of group operations in jth class	56
H	index of translation group	50
H	magnetic field	75
H'_{fi}	matrix element between ψ_f and ψ_i	75
$\mathbf{H}'$	interaction of a radiation field	75
H_{ER}	interaction between electromagnetic radiation and electronic polarizability	104
H_{EL}	interaction between the electron of an electron–hole pair and a phonon	106
H	high frequency mode	124
$H_{EL'}$	interaction between the electron of an electron–hole pair and two phonons	142
H	harmonic interaction	141
H_A	anisotropy field	204
$\mathbf{H}_0$	applied magnetic field in z direction	204
H_E	exchange field	204
H	magnitude of magnetic field	204
$\mathcal{H}$	Hamiltonian operator	10
$\mathcal{H}_R$	Hamiltonian operator leading to correlation field splitting	131
$\mathcal{H}_S$	Hamiltonian operator leading to shape splitting	131
I	intensity	103
I_0	initial intensity	103
k	wave vector for electromagnetic waves	6
k_i	wave vector of incident photon	42
k_f	wave vector of scattered photon	42
k_B	wave vector of elastically scattered photon at the Bragg angle to the incident photon	44
k, k', k'', k'''	integer for different normal coordinates	89, 136
k_i^m	photon wave vector in air	121
K, K_i, K_a, K_b	wave vectors for phonons	5, 143
K_x	wave vector for phonons parallel to crystal surface	31
K_p	vectors into which K is mapped by point group symmetry elements of the space group	55
$[K]$	group of K	55
$\{K\}$	star of K	56
l	integer	10
l	index of group	47
L	low frequency mode	124
L	symmetry point of first Brillouin zone	145
L	length of crystal	172
m	number of molecules in the smallest volume unit cell	2
m'	atomic mass	10
m_1	atomic mass of atom 1	14

Glossary of terms

m^*	mass of conduction band electrons	200
$M(X), M'(X)$	matrix	54
M	dipole moment induced in crystal by electric field of incident radiation	76
M'	M when inelastic scattering occurs	105
M_H	M induced in crystal by magnetic of incident radiation	109
M	magnetic mode	124
M_a	magnetic moment of atom a	204
n	number of atoms per unit cell of smallest volume	21
n, n_1, n_2	integers	49
n'_i	number of internal modes	60
n_i	total number of modes	60
n	refractive index	86
n'	real part of refractive index	103
n_m	refractive index of medium	120
$\bar{n}$	average number of phonons in a lattice vibration	149
$\bar{n}_a$	$\bar{n}$ for phonon a	149
$\bar{n}$	average refractive index	167
n_1	refractive index of surrounding dielectric	178
n_2	refractive index of crystal	178
n^*	number of electrons of charge e	200
N	total number of atoms	21
N_1, N_2, N_3, etc.	order of finite translation group	44
$N(\nu_t)$	number of scattered photons of frequency ν_t per unit cross-sectional area	172
$N(\nu)$	number of incident photons of frequency ν per unit cross-sectional area	172
$\mathbf{p}$	momentum	10
p	integer	35
P, P_1, etc.	point group symmetry element	51
P	polarization of lattice	87
P_V	polarization that occurs when ions are displaced in a lattice vibration	92
P	mode polarized in xz plane	118
P_F	magnitude of transition dipole moment	133
P	exciton and polaron term	196, 199
$\mathscr{P}_k$	projection operator	64
q	charge on an isolated ion	29
q^*	charge on an ion in a lattice	29
q_k	normal mode	62
$q'_{k'k''} \ q''_{k'k''k'''}$	higher order contributions of distortion moments	136
Q_k	normal coordinate	57
r	position vector	76
R, R'	rotation	23
$R, R(\nu)$	reflection coefficient	95
R	distance from source	165
s	atomic orbital	197
Si^x	site group	52
S	space group	52
S'	non-symmorphic space group	53
S''	irreducible representation of space group	56

Glossary of terms

Symbol	Explanation	Page where symbol is introduced
S	mode polarized in xy plane	118
S	surface mode	124
S_{rs}	field propagation tensor	169
S	scattering efficiency	172
$S(\nu)$	density of states at non-localized mode of frequency ν	189
S	exciton and polaron term	196, 199
$\mathscr{S}i_x^x$	site group operator	70
t	time	5
$t_{n_1 n_2 n_3}$	translation element of group	49
$\mathbf{t}$	translation vector	49
$\mathbf{t}^*$	reciprocal translation vector	55
T	translation	23
T	translation group	49
T^s	symmetric representation of translation group	57
T	transformation matrix	55
T	transition probability per unit time	75
T	absolute temperature	95
TE	transverse electric polarization	178
TM	transverse magnetic polarization	178
T_s	superconducting transition temperature	198
$\mathbf{u}$	position vector	8
$\mathbf{u}_t$	position vector for transverse optical mode	87
$\mathbf{u}_l$	position vector for longitudinal optical mode	87
U	unit cell group	52
U_x	element of unit cell group	52
U'	unit cell group of S'	53
U	unit vector in ξ direction	169
$\mathscr{U}_x$	unit cell group operator	58
v	velocity of propagation of a wave	6
v_0	velocity of sound	9
v	number of molecules with no degrees of rotational freedom	59
V	potential energy of crystal	2
V_k	potential energy of individual molecules in crystal	2
V_E	potential energy due to external vibrations of molecules in the crystal	2
V_{Ek}	potential energy due to internal–external vibration coupling	2
V	volume of particle or crystal	165
w	parameter for crystal shapes	123
w	half width of spectral line	137
W	representation of the thickness of a slab	125
W	symmetry point of the first Brillouin zone	145
$x_1, x_2,$ etc.	integer	55
x	Cartesian displacement coordinate	64
x'	sample thickness	103
x	parameter for sample shape	166
X	amount of M that can be removed in an inelastic scattering event.	77

Glossary of terms

X_m	modified reduced mass	101
X	symmetry point of the first Brillouin zone	145
X	energy variable	175
y	Cartesian displacement coordinate	64
z	Cartesian displacement coordinate	64
x	order of group of subgroup	56
α	phonon polarizability tensor	76
α_ξ	component of polarizability tensor in the direction ξ	76
$\overline{\alpha}$	average value of polarizability tensor	171
α	coupling constant for electron lattice interaction	198
β	compressibility	40
β_ξ	component of the first hyperpolarizability tensor in direction ξ	185
β	Bohr magneton	204
β_a	Bohr magneton for atom a	204
γ	damping factor or Grüneisen parameter (frequency independent)	101
γ'	damping factor (frequency dependent)	104
γ''	anisotrophy	172
γ_ξ	component of the second hyperpolarizability tensor in direction ξ	185
δ	phase difference between incident and reflected rays	103
ϵ	dielectric constant	85
ϵ_0	static dielectric constant	88
ϵ_∞	high frequency dielectric constant	89
ϵ_ν	frequency dependent dielectric constant	93
ϵ'_ν	real part of ϵ_ν	102
ϵ''_ν	imaginary part of ϵ_ν	102
η	dispacement of electron gas	200
θ	angle between incident and scattered photons	42
θ_B	Bragg angle	44
θ''	angle between phonon (K) and crystal z axis	111
θ_i	angle between incident beam of electromagnetic radiation and crystal z axis	117
θ'''	angle from polar direction	165
$\kappa(\nu)$	emission coefficient	95
λ	wavelength	5
λ_k	coefficient representing force constant	63
$\partial \mu_d$	dipole moments caused by distortion of an ion by overlap with other ions	29
μ_ξ	component of dipole moment in direction ξ	76
μ'	reduced mass of an ion pair	87
$\overline{\mu}$	average dipole moment	167
$\mu_{\xi s}$	component of dipole moment in direction ξ at the sth crystal site	169

241

Glossary of terms

Symbol	Explanation	Page where symbol is introduced
μ_ξ^d	component of dipole moment in direction ξ associated with defect	189
μ'_{AC}	reduced mass for mixed crystal	193
ν	frequency	5
ν_{ctr}	centro-frequency	40
ν'	phonon frequency	42
ν_f	scattered photon frequency	42
ν_l	longitudinal optic mode frequency	88
ν_t	transverse optic mode frequency	88
ν_0	infrared frequency	89
ν_{k0}	infrared frequency for vibration k	89
$\bar{\nu}_{kt}$	average frequency of all crystal correlation field components	93
ν_{0s}	molecular gas phase vibrational frequency	94
ν_d	frequency calculated for harmonic forces	100
ν_x	frequency of photon or phonon of wave vector k_x or K_x	100
ν_m	Restrastrahl frequency	101
ν_0'	natural vibration frequency of electron–hole pair	105
$\nu_\parallel$	frequency of extraordinary vibration	111
$\nu_\perp$	frequency of ordinary vibration	111
ν_1, ν_2	frequency of extraordinary modes for a biaxial crystal	115
ν_F	Fröhlich frequency	120
ν'	real part of frequency of virtual mode	124
ν''	imaginary part of frequency of virtual mode	124
$\|2\nu''_r\|$	radiative width	137
ν_b	specific frequency	143
ν_i	phonon frequency	143
ν''_o	zero order vibrational frequency	168
ν_0^e	excitation transition frequency	174
ξ	unit vector in direction of polarization of wave	75
ρ, ρ'	density	8, 168
ρ''	degree of depolarization	172
$d\tau$	volume element	76
T	transmission coefficient	94
ϕ	wave function	5
ϕ	phase difference between atoms	35
ϕ_{U_x}	angle of rotation corresponding to symmetry operation U_x	58
χ	character of matrix	55
$\chi'_j(P)$	character of reducible representation	56
$\chi_k(P)$	character of irreducable representation	56
$\chi_{ki}(U_x)$	character of irreducible representation of U_x for initial state	77
$\chi_{kf}(U_x)$	character of irreducible representation of U_x for final state	77
$\chi_k^{n'}(U_x)$	character of irreducible representation of U_x for nth excited state	68
χ	extinction coefficient for absorption	103

Glossary of terms

ψ	wave function	75
ψ_i	wave function of initial state	75
ψ_f	wave function of final state	75
ψ_{ia}	wave function of initial state for phonon a	78
ψ_{fa}	wave function of final state for phonon a	78
ω_{U_x}	number of atoms invariant for unit cell elements U_x	58
$\omega_{U_x}(m)$	number of molecular groups invariant for unit cell elements U_x	59
$\omega_{U_x}(m-v)$	number of molecular groups capable of rotation invariant for unit cell elements U_x	59
Γ	representation	56
Γ_k	irreducible representation	56
Γ_k^i	irreducible representation for initial state	77
Γ_k^f	irreducible representtion for a final state	77
Γ	transmission	95
Γ	symmetry point of first Brillouin zone	145
Ω	expression for frequency of mode	125
$\|2\Delta\Omega''\|$	total width of vibrational band	137

Index

Abelian Group, 47
absorption, 94, 100–4, 118, 125, 127, 129, 149, 156, 165, 174–7, 180–2, 199, 205–7
absorption coefficient, 103, 178
absorption intensity, 99, 101
absorption, metals by, 180
absorption probability, 76
acetylene, 81, 139
acoustic branch, 14, 15, 16
acoustic phonon velocities, 174
acoustic surface modes, 31, 135
acoustic modes, 4, 13, 23, 24, 26, 31, 33, 35, 36–7, 39, 40, 58, 63, 89, 107, 150, 158–9, 174–5, 181, 183, 187, 189, 199, 204–6
activation energy, 199
activation of normally inactive phonons, 208
activities of fundamentals, 77, 78; of multiphonon modes, 78
adenine, 81
adiabatic approximation, 10, 89
adsorbed actoms–additional surface mode due to, 32
Ag^+ ion, 192
Al_2O_3, 113, 166, 197
alcohol, 166
alkali halide crystals, 192–3
alkali halide, lattice dynamics of, 30
alkaline earth halides, lattice dynamics of, 30
ammonium chloride, lattice vibrations in, 158, 159, 160
AlN, 166
angular forces, 30
angular momentum, 197
anharmonic forces and terms, 27, 87, 100–4, 136, 138–40, 144–5, 148, 151–2, 183, 184, 192
anharmonic perturbation, 137
anharmonicity: 'hard force' case, 100; 'soft force' case, 100
anisotropic crystals – lattice dynamics of, 30
anisotropic exchange, 204
anisotropic exchange coupling, 207
anisotropic thermal expansion, 139
anisotropy, 172, 183; crystal of, 111–2; field, 204–5;
annihilation operator, 149

anthracene, 173
9,10-anthraquinone, 81
anti-F centre, 188
anti-Stokes processes, 45, 105, 107, 150, 175, 185
antiferromagnetic, 202–8
antiferromagnetic resonance, 203–5
antisymmetric scattering tensor (Raman), 179
applied fields, 184–5, 196, 202–8
applied magnetic field, 184, 202–8
argon, 100, 189, 191
As_2Se_3, 187
asymmetric vibrations, 183
atomic forces, 111
atomic orbitals, 196
attenuated total reflection, 177–80
attenuation depth of optical surface modes, 31
audio-frequencies, 89
Avogadro's number, 169

$BaTiO_3$, 166, 199
BaN_6, 80
$Ba_2NaNb_5O_{15}$, 82
BaO_2, 166
band modes, 33, 188–94
bands, 195
band shapes, 136–8, 150–6
barium fluoride, 119
Beers–Lambert law, 169
benzene, 81
π-benzenetricarbonyl chromium, 81
benzil, 81
beryl, 197
BH_4^- ion, 192
biaxial, 50;
biaxial crystals, size effects in, 115, 151
binding between states, 145
binding energy, 207
BO_2^- ion, 192
Boltzmann distribution law, 167
Born–von Kármán boundary conditions, 21, 24, 49
Born–Oppenheimer approximations, 10, 159
Born–Mayer potential, 183
boron, 189
Bohr magneton, 204
bound polarons, 193, 199, 200

Index

boundary conditions, 17
Br⁻ ion, 192
BrCN, 170–1
Bragg circle, 44
Bragg conditions, 12
Bragg reflection of X-rays, 12
bromine, 81
Brewster's angle, 178
Brillouin scattering, 107–8, 174–5
Brillouin zone, 207
Brillouin zone boundary, 207
Brillouin zone edge, 31
Brillouin zone (first), 12, 15, 17–19, 43–4, 55–6, 143, 145, 153, 196
Brillouin zone reduced, 17
bulk modes, 31, 117–18, 121–2, 129, 132, 134

$CaMoO_4$, 113, 161–2
$CaWO_4$, 113
cadmium fluoride, 193, 200
cadmium oxide, 200
cadmium sulphide, 119, 183–4
caesium halides, 90, 180, 182
calorimetric absorption measurement, 175
calcite: unit cell group analysis, 60–70, 80; general, 113
calcium fluoride, 89, 119, 146
calcium hydroxide, 145
capacitance measurements, 88
carbon dioxide, 81, 171, 190, 191
carbon disulphide, 81
carbon monoxide, 81, 190
cartesian displacement coordinates, 63, 69
CD_2Cl_2, 81
CeF_3, 80
central forces, 30
centrosymmetric, 162
centre of gravity, 158
centre of inversion, 192
centro frequency, 40
centrosymmetric crystals, 109
CF_3CN, 81
CH_2Cl_2, 81
$(CH_3)_2SO_2$, 81
character, 55–6, 58–60, 64, 77
character tables, 56–7, 77, 83, 209
charge distortion, 189, 191
characteristic absorption, 177
chlorine, 81
chloroform, 166
class, 47
clouds, 165, 198, 199
CO_3^- ion, 60–74, 192
$CoCl_2 \cdot 2H_2O$, 207
CoF_2, 206–7
CoO, 199

coefficient of thermal expansion, 183
ClCN, 170–1
CN⁻ ion, 192
$C_3N_3Cl_3$, 81
coherent scattering of neutrons, 43–4
cold-neutron scattering, 45
colour centres, 188–194
combination bands, 3, 78–9, 82, 94, 102, 139–156, 163, 190–2, 200
combined density of states, 145, 146
compressibility measurements, 26, 40, 89, 101, 183
complex anion, 160
complex crystals, lattice dynamics of, 30
complex frequency, 124
complex oxides, 160
combination maximum, 155
concentric hollow cylinder, 117, 122
conduction band, 106, 195–7, 200
conduction electrons, 30, 197, 201
conservation: of energy, 26, 42–3, 106, 143, 174, 181; of momentum, 26, 181, *see also* conservation of wave vector; of wave vector, 26, 32, 42–3, 106, 143, 147
continuum approximation, 7, 8, 9, 12–14, 85
contributions to character per unshifted atom, 58
convergence errors, 173
coordinates, 57
coordinate, systems, 63
correlation field splitting, 2, 63, 71, 91, 93, 131, 151, 157, 186
correlation table, 71
coset, 47
cosinc modes, 125–8
cotangent modes, 125–8
coupled longitudinal plasmon longitudinal optical phonon, 201
coupling: of internal torsional vibrations and external vibrations, 35, 82, 157–61; of vibrational modes, 157–61, 190; *see also* mixing
Coulomb forces, 28–9, 87, 90–1, 105–6, 116, 133, 195
Coulomb-induced distortion, 92
covalent crystals, 28–39, 79–82, 85, 116–20, 133, 152–6
covalent crystals, examples of spectra of, 79–82
covalent crystal, lattice dynamics of, 28–30
covalent cubic crystals, shape effects in very small crystals, 116–20
Cr^{3+} ion, 197
creation operator, 149

245

Index

critical points, 145–6, 148, 187
crystal of extremely small size, 116–120 with one molecule in the unit cell, 79, 157–61
crystal axes, 111
crystal cylinder, 120, 122
crystal disorder and defects, 32, 187–194
crystal environment, 158
crystal field, 196, 197
crystal orientation, 152, small to large size, 120–31
crystal structure, 46–82, *see throughout the book*
crystal surfaces, 30, 100
Cs^+ ion, 192
$CsMnF_3$, 207
$CsUO_4$, 79
Cu^+ ion, 192
$Cu(thioacetamide)_4Cl$, 80
cubic crystal, 89, 108–110, 116, 158, 159, 183, 184; *see also* individual crystals
cubic ionic crystal cylinder, 120, 122
cubic ionic crystal slab, 117–19, 121–2
crystal ionic crystal sphere, 120, 122
cut off of windows, 182
cyclic boundary, 18, 34, 116
cyclic group, 47
cylohexande d_{12}, 81
cylindrical crystal, 120, 122, 124, 131

Davydov splitting – *see* correlation field splitting
damped oscillator model, 103
damping, 121; due to collisions, 137
damping factor, 101, 104, 136–8, 183, 192, *see also* Grüneisen parameter
dark, 168
de Broglie relation, 26
Debye frequency, 38–9
defects, 3, 32–4, 183, 187–94, 197, 199
deformation, 198–200
deformable molecules, 161
degenerate modes, 67–9, 71, 157, 158, 160, 205
degenerate states, 132, 158, 196
degree of depolarization, 172
density: of radiation, 105; of states of a crystal, 39, 44, 145, 146, 187, 191, 206; of states, two-magnon, 206
depth of penetration, 178
deuterium, 164, 190; halides, 35, 162, 170; ions, 191–2, lattice dynamics of, 30
diagonally cubic crystals, 89
diagrams of normal modes, 64
diamond, 4, 182, 184, 189

diatomic cubic crystals, 31, 134, 135, 182
diatomic crystal, 13–16, 19, 20, 33, 109–15, 117–20, 136, 137
dielectric constant, 85, 87–91, 93, 95, 101–3, 120–1, 123, 125, 127, 129, 131, 134, 174, 178, 200; complex, 102, 103; high frequency, 89, 93; static, 88
dielectric linearity, 101
difference bands, 26, 139–56, 182, 191
1,2-dihydroxy cyclobutenediene, 81
dipolar coupling, 169, 170
dipole-induced dipole intractions, 156
dipole moment, 74–6, 78, 80, 83, 84, 85, 92, 104, 105, 136, 146, 147, 148, 167, 168, 169, 170, 183, 189
direction-dependent phonons, 109–15
discs (sample), 133
disordered crystals, 3, 32–4, 82, 187
distortion, 32, 182–4, 198–200
dislocations, 193
displacement coordinates, 63, 64
dispersion relation, 9, 12–14, 16–17, 20–2, 26, 27, 35–40, 43–4, 92, 98, 99, 108, 114, 122, 145, 152–3, 155, 201, 204, 206
dispersive forces, 28–30, 156
doped crystals (see semiconductors), 199
double potential well, 186, 192
dressed photon, 201
dynamical mechanical properties, 7
dysporsium, 205

effect of applied fields, 184, 192
effect of crystal disorder and defects, 32
effect of finite crystal size, 30, 116–35
effect of polymeric linkages, 34
Einstein frequency, 38
elastic band, 8, 13–14, 22, 26, 40
elastic constants, 7, 175, 177
elastic continuum approximation, *see* continuum approximation
elastic modulii, 7
elastic scattering, 26, 31, 76, 94, 135, 164–6, 174, 185
elastic waves, 7–17, 40
electric-dipole moment, 76
electric-dipole transitions, 76
electric field, 74, 77, 86–8, 90–3, 95, 99, 101, 104–5, 173, 178–80, 184–5, 189, 192, 206; local, 87, 90–5
electric mode, 124, 130–1
electric moment, 87
electrical conductivity, 195
electrical quadrupole, dipole interaction, 206
electrical resistance, 197
electromagnetic field, 75, 95

Index

electromagnetic radiation, 74–9, 87, 88, 95–9, 100, 104–6, 110, 116–17, 131, 174, 180, 185, 195, 201, 206
electron, 75–6, 198–200
electron concentration, 200–1
electron conductivity, 196, 199
electron density, 199
electron gas, 200
electron–hole pair, 104–6
electron irradiation, 189
electron lattice interaction, 198
electron longitudinal phonon coupling, 184
electron motion, 29
electron pairs, 198
electron–phonon interaction, 193
electron scattering, 135
electronegativity differences, 29
electronic coordinates, 10
electronic energy levels, 184, 195–8
electronic polarizability, 104, 105, 108, 109
electronic polaron, 198
electronic Raman effect, 197
electronic states, 75
electronic transitions, 195–200
electrostatic forces, 29, 111, 156
electrostatic interaction, 198
element, 46
elemental crystals, 189, 190, 191
emerald, 197
emission, 94, 149, 175, 199
emission coefficient, 95
'end effects' in polymers, 35
energy exchange, 149
energy gap, 17
energy gap in superconductors, 198
energy loss spectrum, 135
equilibrium site symmetry, 158
ethane, 139
ether, 166
ethylene, 139, 171
Eu^{2+} ion, 192
exchange coupling, 206
exchange field, 204
exchange interaction, 203
exchange interactions, 206
exchange resonance, 203, 204, 205
excitations other than phonons, 104
exciton, 106, 133, 174, 195–8; in paramagnetic species, 196–7; in superconductors, 197–8; polaritons, 109
experimental studies I.R. 133–4; electron scattering, 135
external reflection, 178
external vibrations, 1–2, 4–5, 22, 27, 29, 35, 58–60, 66–70, 74, 82, 93, 145, 158–169, 170–1, 186, 190–2
extinction coefficient, 103, 129, 130, 177–8
extraordinary modes, 111–15

F centre, 188, 193
F_A centre, 188
$F_2, F_3, F_4, F_2^1, F_3^1, F_4^1, F_2^+, F_3^+$ centres, 188
face-centred cubic crystals, 21, 145, 169; see also individual crystals
factor group, 48, 50–4
factor group analysis, 70
factor group splitting, 2
Fe^{2+} ion, 197, 207
FeF_2, 205–7
ferrimagnetic, 202–8
ferimagnetic resonance, 203, 204, 205
ferroelectrics, 139, 152, 184
ferromagnetic, 202–208
field effect, 93, 167, 169
film thickness, 175
finite crystal, 17, 30, 116–35
five-magnon bound states, 207, 208
fixed boundary, 18
force constant, 4, 10, 14, 27, 30, 33–5, 38, 41, 89, 139; anisotropy of, 4
forced wave propagation, 111
formic acid, 81
Franck–Condon principle, 199, 200
free-electron gas, 180
free rotation, 164, 190
Frenkel defect, 188
frequency gap, 17
frequency shifts, 156, 182–4; with pressure, 182; with temperature, 151, 138–40
frequency splittings, 156
fundamental vibrations, 1, 6, 14, 94, 102, 138, 139, 140, 141, 143, 145, 148, 149, 151, 152, 168, 187, 205–8

GaAs, 190
GaSb, 190
gallium phosphide, 109
gap modes, 31, 33, 192–3
garnets, 197
gas-phase, 1, 94, 144, 156–8, 161, 169, 170, 173, 187, 190
Gauss' theorem, 87
general waves in vibrational spectra, 109–15
generating vector of the star, 56
germanium, 180, 189
glasses, 187
glide planes, 49, 53–4
glycine, 113

247

group, 46; of K, 55–6, 78
group theory: application to a crystal lattice, 46; reviews of, 46
group velocity, 9
Grüneisen parameter, 183

H centre, 188
halogens, 41, 81, 171, 173
Hamiltonian, 10
hardness measurements, 101
harmonic approximation, 11, 16, 89, 168
harmonic forces, 11, 16, 27, 32, 100, 200
harmonic terms, 11, 27
Hass–Hornig, equation, 95
heteroatomic systems, 29
hexabromoethane, 81
hexagonal close packed, 164
Hi^-, 188, 192
Hi^0 centre, 188
high frequency dielectric constant, 89
high frequency modes (size), 124
high frequency surface mode, 117, 124–31
high pressure methods, 184, 190
H_2O_2, 186
HOD, 186
hopping motion, 199
host modes, 189, see non-localized modes
hot bands, 151
Huang's theory, 115, 116, 122
hybrid modes–see mixed modes
hydrogen, 164, 190
hydrogen bonding, 185–6
hydrogen chloride, 81
hydrogen halides, 35, 162, 170–1
hydrogen ions, 191, 192; lattice dynamics of, 30
hydrogen sulphide, 81
hydrogen deuterated, 81
hyper Raman effect, 185
hyperpolarizability tensor, 185

I^- ion, 192
I_2Cl_6, 79
ice, 187
identity element, 47
impurities, 156
InSb, 190
incoherent scattering of neutrons, 43
induced absorption, 184–5
induced dipole moment, 76, 169
induced host modes and gap modes, 193
induced modes, 192–3
induced ionic charge, 184
induced Raman scattering, 184–5
inductive forces, 29, 156–7

inelastic scattering, see particular type and Raman and Brillouin scattering; of neutrons, 26, 42–3, 76, 177, 183; of X-rays, 43
inert gas crystals, 189
infinite biaxial crystal, 115
infinite crystal, 7–18, 24
infinite crystal sample, 116, 123
infinite cubic crystal, 109
infinite diatomic cubic lattie, 89–90, 93
infinite ionic crystal, 94–100
infinite polymers, 35
infinite uniaxial crystal, 111–15
in-plane vibrations of polymers, 34
infrared, see throughout the book
infrared intensities, 166
infrared selection rules, 74–9
infrared transmission studies, 175
insulators, 28–30
intensity, 26, 82, 94, 101, 136, 148, 153, 165, 166–73, 174, 175, 179, 183, 187, 189, 191, 193, 197; pressure effects, 3; multiphonon bands, 148, 152, 154; energy, 169; of radiation with a crystal, 83–135
interatomic forces, 26–30, 62
interchange group, 51, 53
interchange element, 51
interference, 94, 133, 151, 164–6, 175
internal reflection, 177, 178, 179
interchain forces, 35
intermolecular forces, 2–4, 26–30, 35, 38, 62, 93, 152, 156–7, 161, 169, 170
intervalence transfers, 199
internal coordinates, 63
internal–external vibration coupling, 2, 145, 156–61
internal mode couplings, 156–8
internal rotation, 161
internal vibrations (modes), 2, 4, 5, 23–4, 67–9, 70, 93, 131, 133, 145, 156–61, 170, 185–6, 191–2
International Tables, 54
intramolecular forces, 2–4, 29, 35, 93, 152
invariant subgroup, 47
inverse power potential, 183
inverse Raman scattering, 185
iodic acid, 155, 156
iodine pentoxide, 81, 153–6
ion-pair, 87
ionic compounds, 156–7, 165, 182, 191–4, 196–7
ionic covalent character, 160
ionic crystals, 85, 133; examples of spectra of, 79–82; defects in, see defects; lattice dynamics of, 28–30

Index

ionic crystal cylinder, 120
ionic crystal sphere, 120
ionic distortion, 92
ionic polarizability, 104, 105
ionicity, 29, 92
ionized polaron, 200
iron pyrite, 80
irreducible representation, 55, 57, 63, 64, 66, 70, 71, 78, 80, 144, 148, 159, 160
irreducible tensor components, 197
irrotational solution, 87, 116
isomorphic group, 47
isomorphic replacement, 160
isotope frequency shifts, 161
isotopic substitution, 157, 160-2, 186
isotropic crystals, 50
isotropic molecule, 77
isotropic solid solutions, 157

$KHCH_2(CO_2)_2$, 80
kinetic energy of a wave, 38
Kirchoff's law, 94
$KMnF_3$, 207
$KMgF_3$, 131-2, 134
$KNiF_3$, 207
$KNiF_4$, 207
KRS-5, 180
KRS-6, 180
Kramers-Kronig relations, 103, 179
krypton, 190
$KTaO_3$, 184
K_2UO_4, 79

$LaCoO_3$, 199
LaF_3, 80
Landau levels, 184
Laporte selection rules, 196
large polaron, 199
laser excitations, 173
lattice constant, 196, 198
lattice dynamics, 116, 187; alkali halides, 30; alkaline earth halides, 30; alkali halide crystals, 30; anisotropic crystals, 30; complex crystals, 30; covalent crystals, 28-30; crystals containing intestitials and vacancies, 33; deuterium, 30; hydrogen, 30; ionic crystals, 28-30; mercurous halide crystals, 30; metals, 28, 30; naphthalene, 161; partially ionic crystals, 29, 30; phonon zig-zag chain, 34; real crystals, 27; simple systems, 7;
lattice vibration, 4, 22, 196; parameters for body and face-centred cubic lattices, 90
lead, 41, 180-1
Li^+ ion, 192
$LiIO_3$, 113-14

$LiYF_4$, 80
librational modes, 22, 161-2
libron, 163-4, 190
libron-libron interaction, 164
line groups, 82
linear chain model for lattice vibrations, 9, 17, 32, 35
linear diatomic lattice, 13
linear elastic band, 8, 40
linear molecules, rotational modes in, 59
linear monatomic lattice, 9, 17
linear multiatomic lattice, 16
linear response theory, 133
lithium halides, 90, 119, 191
lithium hydride, 90
liquid: Resonant Raman effect in, 174
liquid carrier, 166
liquid forces in solution, 156
liquid lattice vibrations, in 3
liquid studies, 161
local electric field, 87, 90-5
local longitudinal polarization field, 198
local symmetry, 70-1; gas modes, 33, 192; mode, 118; modes, 33, 188-94; phonon, 200; two-magnon processes, 207; vibrations, 31
long range forces, 116
longitudinal plasmon, 201
longitudinal stiffness, 8, 12
longitudinal transverse splitting, 2, 21, 83-94,-111-13
longitudinal transverse mixing, 111-12
longitudinal vibrations, 7, 11, 20-1, 83-5, 87-8, 95-8, 107-8, 170, 173, 175-7, 180, 184, 187, 198-201
longitudinal waves, 109, 110, 111-15, 116-18, 121-2, 135
Lorentz sphere, 91
low frequency surface mode, 117, 124-31, 135
Lyddane-Sachs-Teller relation, 88-90, 108

M centre, 188
macroscopic electrostatic field, 85, 87, 108, 109
macroscopic models, 85-90, 101
magnesium fluoride, 113
magnesium hydoxide, 139, 145
magnesium oxide, 145, 166
magnetic dipole interaction, 205
magnetic field, 74, 99, 109, 124, 184, 197, 207, 208
magnetic mode, 124, 129-31
magnetic moment, 202-4
magnon, 202-8

Index

magnon cluster excitations, 207–8
magnon–magnon interactions, 207
maleic anhydride, 81
mass of electron, 198
mass adjusted coordinates, 63–70
matrix-isolated species, 156, 190–1
Maxwell's equations, 75, 87, 96
mechanical energy, 96, 99
mercurous halides, 30
metals: defects in, 32; inelastic neutron scattering by, 45; infrared absorption by, 180–1; lattice dynamics by, 28, 30
methane, 190
microscopic approach to crystal potential energy, 28
microscopic models, 87, 90–3, 101
mixed crystals, 193
mixed modes, 80, 111–15, 132, 144, 150–1, 156, 160, 162–4, 182, 201–2
MnF_2, 192, 207
MnO, 166
molecular crystals, 79–82, 93–4, 133, 139, 152, 156, 190–1; defects in, 32, 189–91; examples of spectra of, 79–82; lattice dynamics of, 28–30
molecular scattering, 104–6, 165
moment of inertia, 162
momentum, 26, 43
monatomic lattice, 9–13, 17–20, 31, 33
monoclinic crystals, 115
mulls, 80–1, 133
multiatomic lattice, 16–17, 21, 112–13, 134
multiorder continuous spectra, 141–2
multimagnon processes, 206–8
multiphonon processes, 78–9, 136–7, 139–56, 182, 186, 199; examples of, 151; infrared mechanism, 139; number of bands due to, 145; temperature dependence of, 148
multiphonon neutron scattering, 43
multiphonon Raman scattering, 141, 185
multiphonon–single phonon mixing, 150–1
multiple paths, 94
mutlipole interactions, 133

N centre, 188
N_2^- ion, 193
Na^+ ion, 192
$NaBF_4$, 80
$NaIO_3$, 80
naphthalene, 81, 161, 173
1,4-naphtholaquinone, 81
$NbOCl_3$, 82
NCO^- ion, 192
NdF_3, 80

near forward Raman scattering, 109, 115, 117–19, 177, 202
neutron inelastic scattering, 26, 43–5, 76, 139, 177; examples of, 43–5; hydrogenous materials of, 42–5; inorganic materials of, 45; organic materials of, 45
neutron irradiation, 189
NH_4^+, 192, see also ammonium compounds
Ni^{2+} ion, 207
NiF_2, 207
Ni_2GeO_4, 161
NiO, 199, 207
Ni (thiourea)$_4Cl_2$, 80
nitrates, 80, 180
nitrous oxide, 81
nitrogen, 81, 139, 171, 190
NO_2^- ion, 192
NO_3^- ion, 192
non-centrosymmetric systems, 162, 184
non-crossing rule, 98, 115, 150
non-diagonally cubic crystals, 89
non-isotopic molecule, 77
non-linear dielectric nature, 102, 136, 148
non-linear molecules, rotational modes in, 59
non-linear Raman effects, 185
non-localized modes, 33, 188–94
non-normal incidence studies, 117–19, 177
non-radiative modes, 120–31
non-symmorphic space group, 52, 57
non-zero wave vector modes, 32, 78, 94, 107, 108, 145, 147–8, 162–4, 174, 183, 187
normal coordinates, 57, 63, 65, 92–3, 136, 167–9, 171–3
normal mode, 21, 24, 46, 57, 62–70, 72–3, 77, 120, 187, 191–2
nuclei, 75
nuclear–nuclear interactions, 28
number of multiphonon modes, 145

O_2^- ion, 192
OCS, 170–1
off-centre behaviour, 192
OH^- ion, 192
one-dimensional lattices, 162, 193; see also elastic waves
one-magnon processes, 206
optical constants, 88–90, 93, 95, 103, 177–8
optical contact, methods of establishing, 180
optical branch, 14–16

250

Index

optical modes, 14–16, 23–4, 26, 36–9, 87, 93, 98, 100, 107, 113, 135, 173, 177, 184, 187, 198–200, 204
optical selection rules, 43
optical surface modes, 31, 135
order of the group, 47, 53
ordinary modes, 111
ordinary–extraordinary ray splitting, 111–15, 151
orientation forces, 29
oriented gas model, 93–4, 157
OsO_4, 81
overtone bands, 78–9, 94, 139–56
overlap distortion, 189, 191
overlap forces, 87, 91, 100
oxygen, 81, 173, 193, 205

P polarized modes, 117–19, 124–5
paramagnetic species, 196, 202
2,2',paracyclophane, 81
parallel vibrations, 111
partially ionic crystals, 29, 30
particle scattering, 164–6
$PbCl_2$, 80
periodic solutions, 88
permanent dipole moment, 169
perpendicular vibrations, 111
petrol, 166
perturbation theory, 106
$PH_4^+Br^-$ ion, 157
phase change in crystal, 139
phase difference between atoms during vibrations, 35–8
phase velocity, 9, 86
phenomenological treatments, 27, 113
phonon, 25–6, 33, 38–9, 42–4, 83–5, 76–9, 96–9, 100, 105–11, 129, 135, 139–56, 180–1, 184, 189, 190, 198–200, 204–6
phonon creation, 149, 150
phonon destruction, 149, 150
phonon frequency distribution, 145
phonon lifetime, 137
phonon occupancy, 149
phonon–phonon interaction, 136–7, 139–56
phonon–photon interaction, 95–9, 100–4, 135
phonon scattering, 193–4
phosphorus, 189
photolysis, 174
photon, 25, 76, 85, 95–9, 100, 106–7, 133, 149, 180–1, 197
photon–exciton interaction, 105–6, 174
piezoelectric crystals, 173
Placzek theory, 171
planar zig-zag chain, 34–5

plane group, 82
plasma frequency, 200–1
plasmariton, 202
plasmon, 200–2
point charge, 29, 93
point defect, 32–4, 188
point dipole, 29
point group, 49, 50, 70, 82–3, 108, 197
polar molecules, 156
polariton, 4, 95–9, 100–9, 114, 120–31, 133, 135, 137–8, 140 149, 174, 201
polariton scattering, 108
polaron, 186, 198–200
polarizability tensor, 77–80, 147, 169, 171–3, 183
polarizable ion, 29, 41
polarization cloud, 198–9
polarization direction of wave, 75
polarization field, 96
polarization lattice, 87–93
polarization of modes, 120, 131
polarized radiation, 77, 105, 152, 172–3, 180
polyatomic compounds, 190
polyglycine, 81
polymers, 3, 34, 82, 157, 160; external vibrations, 82; unit cell group analysis of, 81
polymethylene, 35
polypropylene, 34
polythene, 34, 134, 182
polyvalent cations, 160
position vectors, 10
potassium halides, 90, 100, 134, 192–3
potassium nitrate, 60–70, 80
potential energy, 2, 10, 38, 93, 136, 139, 148, 186, 190–1; barrier to rotation, 161 163–4
powder samples: examples, 79–82, 152; experimental methods, 166; interference 3, 164–6; scattering, 3, 164–6; size effects, 133–5, 151;
precessing spin, 203–5
pressure, effects, 138, 182–4
PrF_3, 80
projection operator, 64
propanoic acid, 81
proton transfer model, 186
proton tunnelling, 186
pseudo-harmonicity, 138, 183–4
pseudo-localized mode, 33
pseudopotential modei, 28
pulse lasers, 175
pure absorption, 175–7
pyridine, 81

Index

quadrupole moments, 30, 169
quantization of lattice vibrations, 25
quantum mechanical treatments, 25, 104
quartz, 113, 150, 175, 182
quasi-harmonicity, 138
quasi-particle, 186

R centre, 188
radiative energy, 96–9
radiative bulk modes, 124
radiative modes, 120–31, 137
radiative surface modes, 120–31
radiative width, 137
radio-frequencies, 89
Rb^+ ion, 192
$RbMnF_3$, 207
$RbUO_4$, 79
Raman, *see throughout the book*
Raman intensities, 171
Raman mechanism, 104–6; for magnons, 205
Raman near forward studies, 4
Raman resonant effect, 174
Raman scattering geometry, 42, 106
Raman selection rules, 74–9
Raman surface mode studies, 135
random polymers, 187
rare earth garnets, 80
rare earth ions, 80, 196–7, 200
rare earth metals, 205
Rayleigh scattering, 31, 174, 185, *see also* elastic scattering
real crystals, 26–38, 101, 116–35
real response, 133
reciprocal lattice vector, 44, 143–4
reduced Brillouin zone, 17
reduced mass, 86, 92, 100–1
reducible representation, 55, 57–9
reflection, 94, 100–4, 118, 125, 127, 151–3, 175, 177–8, 182, 190, 201; infinite ionic crystal of, 94
reflection coefficient, 95, 175, 177–8
reflection losses, 151, 173, 175
reflectivity, *see* reflection coefficient
refractive index, 86, 89, 167, 177, 179; complex, 103
relaxation time, 199
representation of a group, 54
repulsive forces, 100, 133, 145, 156–7
Restrastrahl frequency, 101
restoring force, 200
resonance Raman effect, 174
resonant band mode, *see* mode
resonant mode, 33, 188–194
retardation, 96, 115, 135
rigid ion model, 28–9

ring type molecules, 163
rotation groups, 48
rotation translation coupling, 190
rotational mode, 22–3, 36–7, 39, 59, 63, 66–70, 158–64, 170, 190; of polymers, 36
rotons, 164
rubidium halides, 90

S polarized modes, 117–19, 124–5
S_2^- ion, 193
S_3^- ion, 193
scattering, 133, 151
scattering efficiency, 172–4
scattering maxima, 165
scattering tensor: asymmetric, 197; symmetric, 77
Schönflies system, 48, 54
Schottky defect, 188
screw axes, 49, 52–4
Se_2^- ion, 193
SeS^- ion, 193
second order line spectrum, 141–2
selection rules, 74–9, 108, 148, 174, 185, 197, 206–7
selenium, 81, 113
self-conjugate subgroup, 47
self-energy shift, 138
semiconductors, 28, 190, 201
shape: of crystals, 3, 30–2, 123, 131, 166; *see also* size; splittings, 132, 151
shell models, 29
shifts, 2, 156, 190
short range forces, 28–30, 111, 170
SiF_4, 79, 82
silicates, 80
silicon, 180, 182, 189
silver bromide, 90, 119, 139, 200
silver chloride, 90, 119, 190
single crystal studies, 3, 79–82, 175, 177, 182, 184
site group, 51–7, 70–4, 79, 157–8, 196
site group analysis, 70, 77, 81
site group splitting, 1, 93, 94, 157
site group tables, 73
size effects, 3, 30–2, 82, 94, 110, 116, 120–31, 135, 137–8, 165–6; experimental studies of, 133; uniaxial crystals in, 133;
slab shaped crystals, 124, 135, 138
slow electrons, 135
smallest volume unit cell, 2, 57; *see also* unit cell
small polaron, 199
SnO_2, 80
sodium halides, 4, 89–90, 177, 192

Index

sodium titanate, 139
soft phonons, 139, 150
solid solutions, 186
solenoidal solution, 87, 116
solutions, *see* liquids
sound waves, 9, 13; *see also* acoustic modes
space group, 48–57, 78, 82, 148
specific absorption, 177
specific heat, 26, 38–40, 175
spectroscopic windows, 166, 182
spherical crystal, 120, 122, 124, 129, 133–4
spin, 202–8; orbit interaction, 205; wave, 205
spinels, 160
splittings, 1–4, 151, 156, 190, 193; *see also* correlation, shape, *and* site splitting
$SrTiO_3$, 180, 184, 199
stacking faults, 32
standing wave, 12, 18
star of K, 56, 78–9
static dielectric constant, 88
static field shifting, *see* site group splitting
static mechanical properties, 7
stiffness, 8
stimulated polariton scattering, 109
stimulate Raman scattering, 109, 185
Stokes processes, 45, 105, 107, 150, 175, 185
strain, 8, 139, 182–4, 191
stress, *see* pressure
strontium fluroide, 119
subgroup, 47
sublattice, 87
substitutional defect, 188–94, 196
succinic anhydride, 81
sulphur, 81
sulphur dioxide, 81
superconductors, 197–8
superimposed modes, 167
supertransparency, 182
surface mode, 30–2, 116–35
surface plasmons, 201
symmetric modes, 159, 183
symmetry points of Brillouin zone, 145–6
symmorphic space group, 52

tellurium, 81, 113
temperature dependence; modes, 192, 205; multiphonon modes, 148–150; width and fundamental frequency, 138–9, 148, 151, 156
terbium, 205
tetraatomic lattice, 16, 17
tetracyanoquinodimethane, 81
thallium bromide, 90
thallium chloride, 90
thermal conductivity, 143, 193–4, 196
thermal deformations, 139
thermal neutrons, 43
thermal phonons, 26, 38–40, 102, 107, 148, 151, 193–4
thin films, 118, 175
thin slab limit, 117–19
three dimensional lattice, 20–2, 24, 32, 35, 104, 105, 145, 195
Ti^{3+} ions, 197
TiO_2, 180, 199
tightly bound excitons, 195
Tl^+ ion, 192
tobacco smoke, 165
torsional forces, 30
torsional vibrations, 35, 161
total internal reflection, 120–1
transformation matrix, 55
transition density, 75
transition dipole–transition dipole interactions, 131
transition metal ions, 196–7
transition paramagnetic species in, 196–7
transition probability, 75, 102, 148–9
transition superconductors in, 196–7
transition temperature in superconductors, 198
translation operations, 49
translation group, 49–57
translational modes, 11, 20–2, 36, 39, 59, 63, 66–9, 83–5, 87–8, 96–103, 107–8, 158–64, 170–1, 173, 175–6, 180, 183, 187, 201–2
translational symmetry, 30–2, 124, 187, 192
transmission, 95, 100, 103, 118, 127–8, 151, 165, 175, 178–9, 182
transmission coefficient, 95, 175, 177
transverse electric polarization, 178
transverse longitudinal interaction, 111–15
transvserse magnetic polarization, 178
transverse plasmon, 202
transverse waves, 109–18, 121–3
triclinic crystals, 115
trapped molecule, 190
travelling wave, 9, 11, 19
tunable radiation source, 109
tunnelling spectroscopy, 193
tunnelling transitions, 186, 192
two-dimensional lattice, 32
two-magnon processes, 206–7
two phonon bound state, 145
two-phonon processes, *see* multiphonon processes
Tyndall scattering, 165

Index

U centre, 188, 191–2
u.h.f. devices, 31
ultrasonic methods, 183
ultraviolet radiation, 188, 193, 196, 200
Umklapp processes, 133, 143
uniaxial crystals, 50, 109, 111–15, 180; size effects in, 133–5, 151
unit cell, 2, 5, 13, 16–17, 27, 35, 48, 158, 169, 170
unit cell group, 51–7, 132
unit cell analysis, 57–74, 83–4, 108, 115, 187; activities, 77–9, 109; examples, 79–82
unit cell selection rules, 112
unit cell splitting, 2, 91, see also specific splittings
UO_2, 134
UO_3, 79
uracil, 81

V^{4+} ion, 197
V_K, V_{AK}, V_{KA}, V_F and V_H centres, 188
valence band, 106, 195, 197
valence crystals, 32, see also covalent crystals
van der Waal's forces, 3, 29, 30, 41, 157, 186
vector potential, 75, 105
velocity light, 9, 26, 96
velocity sound, 9, 25
v.h.f. devices, 31
very intense absorption, 103, 151
virtual bulk mode, 120–31
virtual intermediate process, 106, 149, 174, 205–6
virtual modes, 120, 124–5, 137–8
virtual phonons, 206
virtual polariton, 141
virtual response, 133
visible radiation, 109, 188, 193, 196, 199
visible, region, 104
volume, 138

wave model, 4
wave vector, 5, 89, 100, 106, 196; see also zero *and* non-zero wave vector
weakly bound excitons, 196
width: modes, 94, 101, 127, 136–9, 148, 151–6, 185–6, 190, 199; reflection bands, 94, 101; temperature dependence of, 138–9; virtual modes of, 137
window material, 166, 182
WO_3, 79
wurtzite, 113

xenon, 189, 190
XoO_4, 81
X-ray scattering, 12, 26, 60, 82, 170

YF_3, 145
Ytterbium iron garnet, 205
YVO_4, 80

Zeeman effect, 197, 200
zero-point energy, 41
zero wave vector modes, 16, 21, 23–4, 26, 28, 32, 35–6, 56, 70, 77, 79, 83, 85–6, 92, 108–9, 142, 145–8, 150, 163–4, 181, 198, 204–7
zig-zag chains, 35
zinc blende, see zinc sulphide
zinc fluoride, 113
zinc oxide, 135, 183
zinc selenide, 108
zinc sulphide, 92, 108–9, 135, 145, 197
zinc telluride, 108, 119
zone boundary, 12, 40; magnons, 206; modes, see zero wave vector modes
zone edge phonons, 150, 187; see also non-zero wave vector modes